Also by Roger Ellman:

_____   _____

THE ORIGIN AND ITS MEANING

> *ON THE ORIGIN OF THE UNIVERSE*
> *AND ITS MECHANICS,*
> *THE MECHANISM AND ORIGIN*
> *OF INTELLIGENCE,*
> *AND THE IMPLICATIONS*
> *FOR THE INDIVIDUAL AND SOCIETY*

_____   _____

ON THE NATURE OF MATTER

> *THE ORIGIN OF MATTER: ITS CAUSE;*
> *THE STRUCTURE OF MATTER: ITS FORM;*
> *MATTER'S INTERACTIONS: COULOMB, AMPERE, NEWTON,*
> *MATTER WAVES, ATOMIC ORBITAL ELECTRONS,*
> *MATTER AND ATOMS; GRAVITATION;*
> *APPLICATIONS*

_____   _____

THE PHILOSOPHIC PRINCIPLES OF RATIONAL BEING

> *ANALYSIS AND UNDERSTANDING OF*
> *REALITY, TRUTH, GOODNESS, JUSTICE, VIRTUE, BEAUTY,*
> *HAPPINESS, LOVE, HUMAN NATURE, SOCIETY,*
> *GOVERNMENT, EDUCATION, DETERMINISM, FREE WILL, AND*
> *DEATH*

$$v = \mathcal{H}_O \cdot d$$

THE TROUBLE WITH THE HUBBLE LAW

> SIGNIFICANT DATA EVIDENCES THAT THE HUBBLE LAW TREATMENT OF REDSHIFTS IS NOT VALID.
>
> ———
>
> AN ALTERNATIVE TREATMENT OF REDSHIFTS IS EVIDENCED BY FOUR INDEPENDENT OBSERVED COSMOLOGICAL EFFECTS.
>
> THE RESULT GREATLY CHANGES COSMIC DISTANCE MEASUREMENTS AND THE DETERMINATION OF THE AGE OF THE UNIVERSE.

ROGER ELLMAN

Cataloging Data

Ellman, Roger (1932-   )

The Trouble With the Hubble Law
Significant data evidences that the Hubble Law treatment of redshifts is not valid.

THE TROUBLE WITH THE HUBBLE LAW

Copyright (c) Roger Ellman 2018

All rights reserved. This book may not be reproduced nor transmitted in any form nor by any means, electronic, mechanical or other including but not limited to photocopying, recording or by any information storage or retrieval system without written permission from the author, except for the inclusion of brief quotations in a review.

Library of Congress Control Number: 2018903171

Published by: The-Origin Foundation, Inc.,
                1401 Fountaingrove Pkwy.
                Santa Rosa, CA 95403, USA

                707-537-0257

                http://www.The-Origin.org

ISBN 1986570258

CONTENTS

Notes re Author... iii

Preface.. v

Part I - The Hubble Law's Troubles 1

1. The Hubble Law's Troubles 3

Part II - The Universal Exponential Decay 11

2. Analytic Proof: Particles' Central Core and Outward Flow 13

3. The Universal Exponential Decay 19

4. The Universal Decay Redshifts..................................... 25

Part III - The Evidence for the Universal Exponential Decay 31

5. Introduction to the Physical Evidence............................. 33

6. The Galactic Rotation Curves Anomaly 37

7. The Pioneer Anomaly.. 43

8. The Flyby Anomaly 47

9. The Dark Flow Anomaly 55

10. The Common Cause of the Anomalies – The Universal Decay......... 59

11. Calculating The Anomalous Acceleration from The Decay Rate....... 63

Part IV - Conclusion 67

12. Further Experimentally Validating The Universal Decay 69

13. Conclusion 73

14. The Cosmos' Expansion From The Origin To The Present 77

References 113

ABOUT THE AUTHOR

The-Origin Foundation, Inc. is a non-profit organization founded to foster independent scientific, mathematical, and philosophical research.

The author of "The Trouble with the Hubble Law", Roger Ellman, is the General Director of the foundation.

Roger Ellman has published over fifty professional papers on topics ranging from physics, cosmology, and astrophysics to artificial intelligence and mathematics.

He has presented some of his papers to conferences of / at:

>The American Physical Society [APS], .
>The American Society for the Advancement of Science,
>Cambridge University, United Kingdom
>The Library of Alexandria, Egypt
>The Russian Academy of Natural Sciences, St Petersburg
>The Hungarian Academy of Sciences, Budapest
>A SCIRP Science Conference in Shang Hai, China

among others.

He is author of four books in addition to the present "The Trouble With the Hubble Law".

His education includes graduate studies at Stanford University after graduating from West Point, the United States Military Academy.

PREFACE

The successful progress of science depends on and requires the iteration of the following three-step process:

1. The observation of phenomena;
2. The development of potentially valid hypotheses as to the cause and the behavior of those phenomena;
3. The testing of the hypotheses with regard to their validity, generating new, related phenomena;

the cycle repeated until it converges on a fully validated hypothesis.

Of those steps the most difficult is the second, the developing of potentially valid hypotheses, because that is an active creative process of new original material whereas the observation of phenomena and the testing of resulting hypotheses is relatively passive. The first and third steps are dealing with the known. The second step is the creating of new ideas, new knowledge.

For that reason, when new phenomena present the challenge of calling for new, original hypotheses, whether because of the absence of any current hypothesis or because of the failure of all prior attempted hypotheses, the frequent response of the science community at large is to ignore the phenomena, to dismiss them from the scientific search for knowledge.

A few examples of such ignored problems is as follows:

- What makes the stable orbits of atomic electrons to be stable,
- The problem of matter wave frequency,
- How the Universe' first instant came from absolute nothing,

and

- The abundant observational evidence demonstrating the Universal Exponential Decay and that it, not the Hubble Law, is the correct interpretation of redshifts.

PART I -- THE HUBBLE LAW'S TROUBLES

- The Hubble Parameter Measurement Problem.

- The Hubble Law is Asymptotic.

- The Hubble Law Assumptions are not Valid.

- The Hubble Law Requires Exceeding the Speed of Light.

SECTION 1

The Hubble Law's Troubles

COSMIC DISTANCES

The Hubble Law deals with distances from Earth to far distant cosmic objects. In order to accurately treat the subject the method for specifying cosmic distances must be clarified.

The most common popular contemporary unit of cosmic distance is the *light year*, the distance that light travels at velocity $c = 299,792,458.\ m\ s^{-1}$ in one Julian year ($365.25\ days$).

However, at the time of the development of the Hubble Law the unit of cosmic distance was the *parsec* [*pc*], which was based on the length of the *astronomical unit* [*au*].

The *au* was originally the distance from the Earth to the Sun. Because that distance varies due to Earth's orbit being slightly elliptical the *au* was taken as the mean distance. In 2012 the *au* was arbitrarily defined as $au = 149,597,870,700\ m$ (*which is about 150 million km or about 93 million miles*).

The *parsec* was originally defined as the distance at which $1\ au\ subtends\ an\ angle\ of\ 1\ second\ of\ arc$. It was redefined to $pc = [648,000/\pi] \cdot au$ in 2015. That is $1\ pc = 3.086 \cdot 10^{16}\ m$.

The *parsec* is about $3.26\ light\ years$.

THE HUBBLE LAW

The Hubble Law is based on the observation in physical cosmology that the lines in the spectra of various cosmological objects are shifted toward the red of the spectrum, toward longer wavelengths.

- The line spectra on the Earth of the various pure elements exhibit characteristic lines at specific wavelengths from which the element whose spectrum it is can be identified.

- In the line spectra of various observed distant cosmological objects the sets of lines characteristic of specific elements are collectively "reddened", appearing shifted in wavelength toward the red end of the spectrum, longer wavelength, an effect referred to as a "redshift".

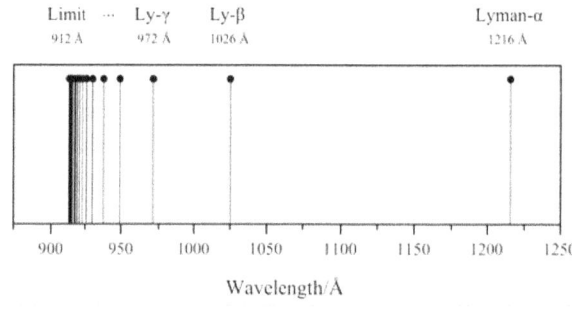

Figure 1-1
Lyman Series of Hydrogen Spectral Lines [Reference: Wikipedia]

In the Hubble Law that redshift is interpreted as a Doppler Effect due to relative velocities of the observed objects away from the Earth [recessional velocities].

The Doppler Effect is the phenomenon that when a source of light is traveling away from an observer he sees the wavelengths of the line spectra of that light shifted toward the red end of the spectrum, toward longer wavelengths. The reason for the phenomenon is that during each cycle of the oscillatory wave form of the light the recession of the light's source away from the direction of its propagation lengthens, so to speak "stretches", the wavelength.

The measured shift, or rather the velocity inferred to be its cause, is found to be approximately proportional to distance from the Earth for objects up to three hundred or so megaparsecs (*mpc*) away per Hubble's then analysis.

The inferred relationship between redshift and distance combined with an inferred relation between recessional velocity and redshift yields a mathematical expression for Hubble's Law as follows:

(1-1) $v = H_0 \cdot d$ or $d = v/H_0$ or $d \approx f(redshift)/H_0$

where:

v is the velocity away from the Earth of the astronomical source [recessional velocity] expressed in km/s.

H_0 is Hubble's constant expressed in km/s per mpc

d is the distance of the astronomical source expressed in mpc

$f(redshift)$ is the velocity Doppler-related to the redshift

THE SEVERAL PROBLEMS WITH THE HUBBLE LAW

1-Hubble Parameter Measurements

The first, and most obvious, problem with the Hubble Law is that of determining the value of its constant H_0. Recent measurements have resulted in drastically different values for the constant. Evidence from the Hubble telescope has produced a value for the Hubble constant of $H_0 = 73$. A recent analysis by a team tracking the problem puts the constant in the range of $72-75$. Using another entirely different method a probable range of only $66-68$ results.

Such results should lead to questioning at least whether H_0 is a true physical constant and more reasonably to questioning the overall general validity of Hubble's Law. Figure 1-2 below lists the numerous various results of attempts to pin down the value of the Hubble constant this 21st century.

Date published	Hubble constant (km/s)/Mpc	Observer
2017-10-16	70.0+12.0 −8.0	The LIGO Scientific Collaboration and The Virgo Collaboration
2016-11-22	71.9+2.4 −3.0	Hubble Space Telescope
2016-07-13	67.6+0.7 −0.6	SDSS-III Baryon Oscillation Spectroscopic Survey
2016-05-17	73.24±1.74	Hubble Space Telescope
2015-02	67.74±0.46	Planck Mission
2013-10-01	74.4±3.0	Cosmicflows-2
2013-03-21	67.80±0.77	Planck Mission
2012-12-20	69.32±0.80	WMAP (9-years)
2010	70.4+1.3 −1.4	WMAP (7-years), combined with other measurements.
2010	71.0±2.5	WMAP only (7-years).
2009-02	70.1±1.3	WMAP (5-years). combined with other measurements.
2009-02	71.9+2.6 −2.7	WMAP only (5-years)
2007	70.4+1.5 −1.6	WMAP (3-years)
2006-08	77.6+14.9 −12.5	Chandra X-ray Observatory
2001-05	72±8	Hubble Space Telescope

Figure 1-2
21st Century Measurements of the Hubble Constant
[Reference: Wikipedia]

2-The Hubble Law is Asymptotic

The relativistic relation between redshift and theoretical recessional velocity for z > 1 is the Fizeau-Doppler formula equation *(1-2)*.

$$(1-2) \quad z = \frac{\lambda_{observed}}{\lambda_{emitted}} - 1 = \sqrt{\frac{1 + v/c}{1 - v/c}} - 1$$

That solved for v is equation *(1-3)*.

$$(1-3) \qquad v = \frac{(z+1)^2 - 1}{(z+1)^2 + 1} \cdot c$$

Equation *(1-1)* with substituting for *v* in it from equation *(1-3)* yields the asymptotic relationship of separation distance expressed in megaparsecs versus redshift, equation *(1-4)*. Since *1 parsec* is about *3.26 light years* the distance, *d*, can be expressed as time into the past that the locally observed light was emitted by its far distant source.

$$(1-4) \qquad d = \frac{v}{H_0} = \frac{c}{H_0} \cdot \frac{(z+1)^2 - 1}{(z+1)^2 + 1} \quad \text{in mpc or} \times 3.26 \text{ in mega light years}$$

Figure 1-3 is a plot of equation *(1-4)* for several different values of the Hubble Constant, values consistent with the Hubble Constant measurements problems presented above.

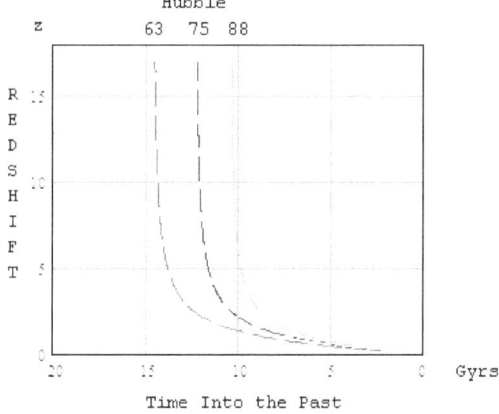

Figure 1-3
The Asymptotic Hubble Law

Unfortunately, the currently more favored value for the Hubble Constant, $H_0 = 73$ fails to correspond well to the currently favored age of the universe of about *13.8 Gyrs* for which the Hubble Constant is about $H_0 = 71.3$.

3-The Hubble Law Assumptions

There are two assumptions on which the Hubble Law is dependent. Neither is supported by facts.

The First Assumption

- It is assumed that the recession velocity of distant cosmic objects is approximately proportional to the object's distance from the Earth.

Figure 1-4, below demonstrates that the assumption that recession velocity of distant cosmic objects is approximately proportional to their distance is sufficiently invalid to nullify the concept of H_0 being a "constant".

That defect alone is sufficient to account for the range of measured values for H_0 given in Figure 1-2.

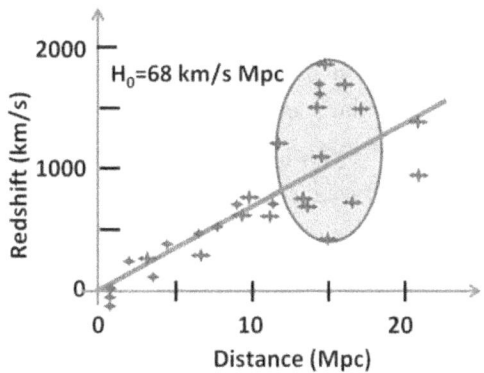

Figure 1-3
Fit of Some Redshift Velocities to Hubble's Law
[Reference: Wikipedia, "Hubble's Law"]

The Second Assumption

- It is assumed that the Doppler Effect is the cause of redshifts.

The quantification of this effect is as follows.

(1-2)
$$f_{observed} = \frac{c}{c+v} \cdot f_{source} \qquad c = \text{light speed} \quad v = \text{recession velocity}$$

$$\lambda_{observed} = \frac{c}{f_{observed}} = \frac{c+v}{f_{source}} \qquad f = \text{frequency} \quad \lambda = \text{wavelength}$$

$$v = \lambda_{obs} \cdot f_{src} - c = [\lambda_{src} + \Delta\lambda] f_{src} - c = \Delta\lambda \cdot f_{src} = \frac{\Delta\lambda}{\lambda_{source}} \cdot c$$

The variable designation for redshift is z per equation *(1-3)*.

(1-3)
$$z = \frac{\lambda_{observed} - \lambda_{source}}{\lambda_{source}} = \frac{\Delta\lambda}{\lambda_{source}}$$

That is z is the ratio of the change or "shift" in the wavelength to the originating light source wavelength. Therefore the recession velocity of equation *(1-2)* is the redshift fraction of equation *(1-3)* of the speed of light as equation *(1-4)*.

(1-4) $\quad v = z \cdot c$

The speed of light, c, is an absolute maximum speed limit. Therefore, $v > c$ is not possible which means that the redshift cannot be greater than *1.0*.

On the other hand many actual redshifts measure to be greater than *1.0*, the largest recently reported being $z = 11.9$.

The discrepancy between the speed of light limit and the high measured values of z is deemed resolved as follows.

It is contended that of the total recession velocity producing the extreme redshift most of that motion is due to expansion of space, expansion of the Universe, not actual classical motion of the receding object. Due to the expansion the distance to remote galaxies can increase at more than light speed, $c = 3 \cdot 10^8$ m/s. It is contended that this does not imply that the galaxies move faster than the speed of light.

What is being contended is that there is a means by which distant cosmic objects can increase their distance from us such that it is as if, for redshift purposes, their speed exceeds light speed while the objects are actually not at all physically moving faster than light.

The objects are said to have their location in space carried along, by expansion of space, at in excess of light speed, but their position in their location is only moving at less than light speed producing a minor Doppler-like redshift component of the overall total redshift due to the spatial expansion.

Substantial development of the concept and substantial mathematics treating the assumed expansion of space describe the effect. But, they do not prove or validate what the equations describe.

Basically the contention that distant objects can get farther from us at a rate producing redshifts as if the object were moving faster than light, but without ever exceeding the speed of light is untenable.

That leaves the Hubble Law in "limbo". There must be an alternative explanation of redshifts that does not defy reason, does not defy logic, that meets the classic requirement of Occam's Razor: the simplest solution is most likely the correct one.

There is such a solution:

The Universal Exponential Decay,

Which follows.

PART II -- THE UNIVERSAL EXPONENTIAL DECAY

- *Analytic Proof: Particles' Central Core
  and Outward Flow*

- *The Universal Exponential Decay*

- *Universal Decay Redshifts*

SECTION 2

Particles' Central Core and Outward Flow

THE PARTICLE "CORE"

Each of the particles of which atoms of matter are made: protons, neutrons and electrons, has at its center a minute dense "core" as follows.

Gravitational Equivalent Frequency

Consider a small individual particle such as a proton or a neutron. The gravitational action of a massive body is the collective effect of the individual action, below, in each of its such particle components.

Newton's law of gravitation expressed in terms of m_{source} and $m_{acted\text{-}on}$ and with both sides of the equation divided by $m_{acted\text{-}on}$ is, of course,

$$(2\text{-}1) \quad a_{grav} = G \cdot \left[\frac{m_{source}}{d^2} \right]$$

stating that gravitation is a property of a body's mass.

However, mass and energy are equivalent, so that a mass, m, is proportional to a frequency, f, that is characteristic of that mass. That is

$$(2\text{-}2) \quad m \cdot c^2 = h \cdot f \quad \text{or} \quad f = [c^2/h] \cdot m$$

so that the m_{source} of equation *(2-1)* has a corresponding equivalent frequency, f_{source}.

That being the case, the gravitational acceleration, a_{grav}, can be expressed in terms of that frequency as the change, Δv, in the velocity, v, of the attracted mass per time period, T_{source}, of the oscillation at the corresponding frequency, f_{source}, as follows.

$$(2\text{-}3) \quad a_{grav} = \Delta v / T_{source} = \Delta v \cdot f_{source}$$

Gravitation and the Planck Length

It can then be reasoned using equation *(2-3)* = equation *(2-1)* as follows.

$$(2\text{-}4) \quad a_{grav} = \Delta v \cdot f_{source} = G \cdot \left[\frac{m_{source}}{d^2} \right]$$

Equation *(2-5)*, below, is obtained by using that frequency is proportional to mass so with f_p and m_p as the proton frequency and mass then $f_{source} = [m_{source} / m_p] \cdot f_p$.

$(2\text{-}5)$ $\quad \Delta v \cdot \left[\dfrac{m_{source}}{m_p}\right] \cdot f_p = G \cdot \left[\dfrac{m_{source}}{d^2}\right]$

Rearranging and canceling m_{source} on both sides of the equation,

$(2\text{-}6)$ $\quad \Delta v = \dfrac{G \cdot m_p}{d^2 \cdot f_p}$ per cycle of f_{source}.

Then substituting, per equation $(2\text{-}2)$, $m_p = [h \cdot f_p]/c^2$,

$(2\text{-}7)$ $\quad \Delta v = \left[\dfrac{G}{d^2 \cdot f_p}\right] \cdot \left[\dfrac{h \cdot f_p}{c^2}\right]$

$\quad = \dfrac{G \cdot h}{d^2 \cdot c^2}$ per cycle of f_{source}.

The Planck Length, l_P, is defined as

$(2\text{-}8)$ $\quad l_P \equiv \left[\dfrac{h \cdot G}{2\pi \cdot c^3}\right]^{\frac{1}{2}}$ so that $G = \left[\dfrac{2\pi \cdot c^3 \cdot l_P^2}{h}\right]$

Substituting G as a function of the Planck Length from equation $(2\text{-}8)$ into G as it is in equation $(2\text{-}7)$, the following is obtained.

$(2\text{-}9)$ $\quad \Delta v = \left[\dfrac{2\pi \cdot c^3 \cdot l_P^2}{h}\right] \cdot \left[\dfrac{h}{d^2 \cdot c^2}\right]$

$\quad = c \cdot \dfrac{2\pi \cdot l_P^2}{d^2}$ per cycle of f_{source}.

This result states that:

- the velocity change due to gravitation, Δv,
- per cycle of the attracting mass's equivalent frequency, f_{source},
   which quantity, $\Delta v \cdot f_{source}$, is the gravitational acceleration, a_{grav},
- is a specific fraction of the speed of light, c, namely the ratio of:
   - 2π times the Planck Length squared, $2\pi \cdot l_P^2$, to
   - the squared separation distance of the masses, d^2.

That squared ratio is, of course, the usual inverse square behavior.

This also means that at distance $d = \sqrt{2\pi} \cdot l_P$ from the center of the source, attracting mass, the acceleration, Δv, per cycle of that attracting mass's equivalent frequency, f_{source}, is equal to the full speed of light, c, the most that it is possible to be. In other words, at that [quite close] distance from the source mass the maximum possible gravitational acceleration occurs. That is the significance, the physical meaning, of l_P or, rather, of $\sqrt{2\pi} \cdot l_P$.

The physical significance of $\sqrt{2\pi} \cdot l_P$ is that it sets a limit on the minimum separation distance in gravitational interactions and it implies that a "core" of that radius is at the center of fundamental particles having rest mass. That is, equation *(2-11)* clearly implies that it is not possible for a particle having rest mass to be approached closer than that distance.

That physical significance of $\sqrt{2\pi} \cdot l_P$, is so fundamental to gravitation and apparently to particle structure, that it more truly represents a fundamental constant than does l_P. For those reasons that length should replace l_P as a fundamental constant of nature as follows.

(2-10)   The fundamental distance constant, δ

$$\delta^2 \equiv 2\pi \cdot l_P^2$$

$$\delta = 4.051,34 \times 10^{-35} \text{ meters} \quad\quad \text{[2006 CODATA Bulletin]}$$

Equation *(2-11)* then becomes equation *(2-13)*.

(2-11) $\Delta v = c \cdot \dfrac{\delta^2}{d^2}$ per cycle of f_{source}

a quite pure and precise statement of gravitation, that gravitation is a function of the speed of light, c, and the inverse square law, in the context of the oscillation frequency, f_{source}, corresponding to the attracting, source body's mass. It makes clear that an oscillation is an integral part of gravitation

THE PARTICLE'S

There being a need for each gravitationally acting [attract**ing**] particle to communicate to each gravitationally acting [attract**ed**] particle the direction from the attract**ed** particle to the attract**ing** one and the magnitude of the attract**ing** particle's gravitational attraction, there must be something flowing, continuously, carrying that information, spherically outward, from every gravitating particle to every other gravitating particle. That flow-communication is the gravitational field, an active process not a static state.

Furthermore, the necessity for gravitation that an oscillation and its frequency are closely involved in the effect [equations *(2-9)* and *(2-11)*] and therefore in what is communicated by the flow means that the flow itself must be oscillatory.

For such a flow to persist and to have persisted the billions of years since the "Big Bang" there must be a supply of that outward flowing substance in every particle. And, that "supply" must be an extremely concentrated reservoir of that which flows outward [concentrated relative to the outward flow].

Having now just determined:

- That δ sets a limit on the minimum separation distance in gravitational interactions and therefore implies that a "core" of that radius is at the center of fundamental particles;
- That there must be something flowing, continuously from every gravitating particle to every other gravitating particle, that flow being oscillatory;
- And that an extremely concentrated reservoir supply of that which is flowing outward is required at the center of each particle to support the billions of years of the outward flow;

The only reasonable conclusion is that that reservoir is a spherical *core* of radius δ at the center of all particles;

And that its impenetrability is due to its immense concentration [billions of years worth of flow] of the flow substance [hereinafter *medium*] in the extremely minute [$\delta = 4.05134 \times 10^{-35}$ meters per equation *(2-10)*] radius *core* at the center of every particle having rest mass.

But, what "contains" that *core*'s supply or why doesn't it all just quickly "slosh" out and be gone ? The answer is that it is trying to do just that, to "slosh" out, as hard as it can. It cannot help propagating outward because it has no container. But it can only propagate outward at the limiting rate determined by its surface area, $4 \cdot \pi \cdot \delta^2$ and the fastest speed possible for it to flow, the speed of light, c.

Having established the supply of medium flow and its on-going outward flow serving the role of gravitational field as a property of every particle exhibiting rest mass, the question arises, "What of the electric field, much stronger than gravitation and co-present with gravitational field whenever the gravitating particle has electric charge ?"

Just as is the case for gravitation every particle having electric charge has a need to communicate its "message" to every other such particle: the direction back from the encountered particle to the transmitting one, the magnitude of the transmitting particles' charge, and the nature of the charge, whether positive or negative. That flow-communication is the electric field, an active process not a static state.

Furthermore, for that flow to persist and to have persisted the billions of years since the "Big Bang" there must be a supply of that which is flowing outward for it at the center of every particle. And, that "supply" must be an extremely concentrated [relative to the outward flow] reservoir of that which flows outward.

Two such simultaneous flows constituting the two fields, gravitational and electric, and two supporting reservoirs supplying the flows is clearly untenable. There can only be one reservoir in each particle's *core* and one resulting flow producing both the gravitational field and the electric field if for no other reason than because two supply reservoirs would mutually interfere with comprehensive spherically outward flow of each.

The process of that *Propagated Outward Flow* gradually depleting the *medium* remaining in the *core*, as that which is flowing and its supply in the *core* will be referred to, is:

The Universal Exponential Decay as in Section 3.

SECTION 3

The Universal Exponential Decay

Since the "Big Bang" the *Propagated Outward Flow* has been gradually depleting the original supply of *medium* in the *core* of each particle. That process, an original quantity gradually depleted by flow away of some of the remaining quantity, is an exponential decay.

THE NATURE OF THE DECAY

Of the three fundamental dimensions of length $[L]$, mass $[M]$, and time $[T]$ only length can decay. Time being the independent variable of material reality, whether it decays, varies, or is rigorously constant is beyond our ability to detect. Likewise, mass cannot decay, it being proportional to frequency, the inverse of time. The dimension that is decaying is length, the $[L]$ dimension in the dimensions of, for example: the Planck Constant, h, $[M \cdot L^2/T]$; the speed of light, c, $[L/T]$; and the Newtonian Gravitational Constant, G, $[L^3/M \cdot T^2]$. The decay process involves the fundamental constants (c, q, G, h, etc.) and decay of any of those must be dimensionally consistent with the decay of the others.

The *core*'s outer boundary is a surface of area $4 \cdot \pi \cdot \delta^2$; but, it lacks the power to restrain or contain anything. It is only a boundary. However, the only way the content of the *core* can leave and flow outward is through the *core*'s surface. That flow is subject to the speed limit of light speed, c. That sets the flow at $[4 \cdot \pi \cdot \delta^2] \cdot c$.

THE RATE OF THE DECAY

A process which the *core* decay resembles is the pumping of gas out of a chamber to create a vacuum. In this case the "gas" is the *medium*, the chamber is the *core*, and the pumping is the loss of *medium*, through the surface boundary of the *core*, to outward propagation. The process of the pumping, whether of gas out of a vacuum chamber or *medium* out of the *core* is such that:

· The rate of change of the amount of *medium* remaining in the *core* equals

· The amount per volume of *medium* remaining, times

· The pumping speed, that is the volume per time of the propagation.

This is based on the conceptualization of the process as:

- The *medium* is uniformly distributed in the *core*;
- A minute increment of volume is then pumped out in a minute time;
- The remaining *medium* then redistributes uniformly within the *core*,
- and the cycle repeats over and over.

From this the rate of change of the amount of *medium* present within the *core* is as follows.

(3-1)
$$\text{Medium Rate of Change} = -\begin{bmatrix}\text{Amount} \\ \text{per} \\ \text{Volume}\end{bmatrix} \times \left[\begin{bmatrix}\text{Pumping} \\ \text{Speed}\end{bmatrix} = \begin{bmatrix}\text{Surface} \\ \text{of Core}\end{bmatrix} \times \begin{bmatrix}\text{Flow} \\ \text{Speed}\end{bmatrix}\right]$$

$$\frac{d\upsilon}{dt} = -\frac{\upsilon}{\frac{4}{3}\pi\cdot\delta^3} \times \left[\left[4\cdot\pi\cdot\delta^2\right]\times[c]\right] = -\frac{3\cdot c}{\delta}\cdot\upsilon$$

The pumping takes place over the entire surface of the *core* and the rate at which the outward flow takes place is the speed of *medium* travel, the speed of light, *c*. [Both *c* and *δ* are functions of time, each decaying in its dimensional unit *[L]*; however, their decay rates are identical so that their ratio, as in equation *(3-1)* is constant.]

Therefore, rearranging equation *(3-1)* and integrating:

(3-2)
$$\frac{d\upsilon}{\upsilon} = -\frac{3\cdot c}{\delta}\cdot dt$$

$$\ln(\upsilon) = -\frac{3\cdot c}{\delta}t + C$$

$$\upsilon(t) = \upsilon_0 \cdot \varepsilon^{-\frac{3\cdot c}{\delta}\cdot t} \quad [\varepsilon^C \text{ evaluated as } \upsilon_0]$$

Therefore, the decay time constant, τ is

(3-3)
$$\tau = \frac{\delta}{3\cdot c}$$

However, that result cannot be correct. Equation *(3-3)* yields a value of about $4.5\cdot 10^{-44}$ seconds. That is completely inconsistent with the universe having an already accomplished life time of billions of years.

It must be concluded that *medium* empties from the *core* at only a minute amount of the volumetric pumping speed used above or, alternatively, that the *core* volume contains, as *medium*, an immense supply of volume, of "highly concentrated volume" so to speak.

In Section 2, *Particle's Central Core and Outward Flow*, under the subtitle "The Particle's Flow", it was stated:

> "For such a flow to persist there must be a supply of that outward flowing substance in every particle. And, for that flow to have persisted the billions of years since the "Big Bang" that "supply" must be an extremely concentrated reservoir of that which flows outward."

However thought of, it must be from the foregoing that an additional factor that reduces the rate of change of the *core medium* must be used in equation *(3-3)* so that it becomes

(3-4) $$\begin{bmatrix} \text{Medium} \\ \text{Rate of} \\ \text{Change} \end{bmatrix} = - \begin{bmatrix} \text{Amount} \\ \text{per} \\ \text{Volume} \end{bmatrix} \times \left[\begin{bmatrix} \text{Pumping} \\ \text{Speed} \end{bmatrix} = \begin{bmatrix} \text{Surface} \\ \text{of Core} \end{bmatrix} \times \begin{bmatrix} \text{Flow} \\ \text{Speed} \end{bmatrix} \right] \cdot \begin{bmatrix} \text{Concentration} \\ \text{Factor} \end{bmatrix}$$

$$\frac{d\upsilon}{dt} = -\frac{\upsilon}{\frac{4}{3}\pi\cdot\delta^3} \times \left[\left[4\cdot\pi\cdot\delta^2\right] \times c\right] \cdot \frac{1}{F} = -\frac{3\cdot c}{\delta\cdot F}\cdot\upsilon$$

where F is the additional factor. Equations *(3-2)* and *(3-3)* then become *(3-5)* and *(3-6)* as follows.

(3-5)
$$\frac{d\upsilon}{\upsilon} = -\frac{3\cdot c}{\delta\cdot F}\cdot dt$$

$$\ln(\upsilon) = -\frac{3\cdot c}{\delta\cdot F}\cdot t + C$$

$$\upsilon(t) = \upsilon_0 \cdot \varepsilon^{-\frac{3\cdot c}{\delta\cdot F}\cdot t} \quad [\varepsilon^C \text{ evaluated as } \upsilon_0]$$

Therefore, the decay time constant, τ is

(3-6)
$$\tau = \frac{\delta\cdot F}{3\cdot c}$$

and the Universal Exponential Decay is

(3-7)
$$\upsilon(t) = \upsilon_0 \cdot \varepsilon^{-t/\tau}$$

The values of δ and c are known, but what is the value of F?

The Particle Core and the Core's Outward Flow

The outward flow is flow of a highly effective substance in that it produces the effects of gravitational field and electric field; yet, it is at the same time a flow of an extremely intangible substance producing only the intangible gravitational field and electric field. But that flow is an integral component of humans' physical world and acts in accordance with well established physical laws dealing with well known physical quantities.

But, the interior of the reservoir supply of *medium* is entirely foreign to us. We can not really conceive of gravitational plus electric field, stationary in place and so dense and concentrated that it supports outward flow of our world's fields over many billions of years, any more than we can truly conceive of infinity.

The only thing known about the *core*, the "Core Domain" as compared to our "World Domain" is that as perceived from outside, in terms of the "World Domain" it appears to be a volume of $^4/_3 \cdot \pi \cdot \delta^3$ with a surface of $4\cdot\pi\cdot\delta^2$.

But, its on-going flow generates potential energy field which in our "World Domain" appears to be static, not flowing. Flowing *medium* is static potential energy field. Thus the *core medium* supplies the *propagated outward flow* potential energy. But, the *core medium* is also the electric charge of the particle.

The dimension units of the *core medium* charge are $[M\cdot L]$ per equation 3-8.

$(3-8)$ $\quad$ {Force} = {[Mass]·[Acceleration]} = $\dfrac{M \cdot L}{T^2}$ Newton's Law

$$\{\text{Force}\} = \frac{\{c^2 \cdot q^2\}}{4 \cdot \pi \cdot L^2} = \frac{M \cdot L}{T^2} \text{ Coulomb's Law}$$

$$\{c^2 \cdot q^2\} = \frac{M \cdot L^3}{T^2} \quad \therefore \{q^2\} = M \cdot L$$

Now for the energy aspect: dimensionally Planck's constant, h, $[M \cdot L^2/T]$ appears as "energy per time" $[M \cdot L^2/T] \cdot [1/T]$ in the relation $W = h \cdot f$ [W is energy and f is frequency]. That oscillation energy, the equivalent of matter energy as in $h \cdot f = m \cdot c^2$, cannot be in the "f" nor its equivalent "$1/T$"; the energy of $W = h \cdot f$ must be in h. The energy potential of the *core*'s *medium* supply is in the Planck Constant. It is in potential form h/c which is of the same dimensional units $[M \cdot L]$ as found in equation (3-8) because the two aspects, charge and *core medium* potential energy are aspects of the same single substance, the *core medium*.

The gradually decaying *medium* contained within the *core* is not merely the geometric *core* physical volume as viewed from our world; it is "highly concentrated volume", the capability if freed into space outside the *core* to be myriad *core* physical volumes, the volume of space.

That difference distinguishes the physics of the *core*'s internal "Core Domain" vs. the outside "World Domain". The ratio to the world view geometrical volume of that "*highly concentrated volume*" of medium to be propagated is the above concentration factor F.

$(3-9)$ $\quad F = \dfrac{\text{Equivalent of Core Medium Supply}}{\text{Geometric Core Volume}}$

$$= \frac{h/c}{\tfrac{4}{3} \cdot \pi \cdot \delta^3} \; \frac{\text{Units } [M \cdot L]}{\text{Units } [L^3]} = 7{,}938{,}010{,}000 \cdot 10^{60}$$

F is a pure number just as are $\tfrac{4}{3}$ and π of equation $(3-8)$. Saying the *core* is *medium* $[M \cdot L]$ vs. volume $[L^3]$ is like saying a year is [days] vs. [seconds].

The factor F spans two different regimes of material reality:

$\quad$ 1 - The natural world regime in which we exist and function;

$\quad$ 2 - The interior of the *core* of each particle, the supply of highly concentrated *medium*, minute portions of which are propagated outward in each cycle of the particle's oscillation, gradually depleting the supply.

From equation $(3-6)$ with the value for F of equation $(3-8)$ the value of τ, the universal decay time constant is

$(3-10)$ $\quad \tau = 3.57532 \cdot 10^{17}$ seconds

$\quad\quad\quad\quad \approx 11.3373 \cdot 10^9$ years

That value of τ and the equation $(3-9)$ value of F are validated by their correlation with the Pioneer Anomaly of Section 7.

This Universal Exponential Decay produces redshifts,

as presented in the following Section 4,

that can extend to very large values of z

without depending on motion of the cosmic objects,

without the problem of the limitation of the speed of light.

SECTION 4

The Universal Decay Redshifts

THE PROPAGATION OF MEDIUM

The *Propagated Outward Flow* of *medium*, which flow carries and travels like light, is controlled by the pair of parameters μ and ε. The situation is the same as that of propagation of electrical signals along a transmission line.

A transmission line is an electrical device for transmitting oscillatory electrical signals from one place to another. Examples are: the various coaxial cables and two-wire pairs found in radio, and video systems interconnecting equipment components. When electrical signals are introduced at one end of such a line they do not appear at the far end instantaneously. Rather, there is a finite speed of travel of the electrical effects along the line.

The reason is that any such line inevitably has some series electrical inductance and shunt capacitance whether intentionally placed there or not. These are distributed along the length of the transmission line in minute increments. The speed of travel of the electrical signal along the line is limited by the time that it takes the signal to build up its full value successively through the electrical inductance of each minute increment and the time that it takes to build up its full value successively on the capacitance of each minute increment.

The electrical inductance of a coil of wire is μ_0 times the dimensions of the coil, $N \cdot A/L$, where N is the number of turns in the coil, A is the cross-sectional area of the coil and L is the length of the coil. The N is a dimensionless number. Thus the inductance is, dimensionally, μ_0 times an area divided by a length, that is times a net length. The μ_0 is then inductance-per-length.

The electrical capacitance of a simple parallel plate capacitor is ε_0 times the dimensions of the capacitor, A/L, where A is the area of each of the two identical plates and L is the distance between them. Thus the capacitance is, dimensionally, ε_0 times an area divided by a length, that is times a net length. The ε_0 is then capacitance-per length.

THE UNIVERSAL EXPONENTIAL DECAY REDSHIFT

The Big Bang took place in absolute nothing. Before the Big Bang there was only absolute nothingness. There was no "free space" with μ_0 and ε_0. After the Big Bang's explosion into all of the particles of the universe, each of those particles was sending its own *Propagated Outward Flow* into nothing, into emptiness.

Where did the oscillatory *Propagated Outward Flow*'s μ_0 and ε_0 come from? The only thing they could have come from was the *Propagated Outward Flow* itself, that is from each particle's *core* medium content. There is no other possible source because everything else was absolute nothing, "the zero of existence". The μ_0 and ε_0 are inherent in the substance of the oscillation, the *medium*, which means, μ_0 and ε_0 are also inherent in the outward propagation. Each particle's *Propagated Outward Flow* contains, carries within it, has its speed of propagation determined by and set at its value at the moment it was emitted, by its own μ_0 and ε_0.

In the propagating medium its μ_0 parameter is its inductance and its ε_0 parameter is its capacitance. Their effect in determining the speed of propagation is the same as in a transmission line, above. The dimensions of each include the factor $1/Length$. As all length aspects decay so do μ_0 and ε_0 consequently "anti-decay". That conforms to their role in the denominator in the speed of light equation *(4-1)*, c decaying in its L/T dimension.

(4-1)
$$c = \frac{1}{\sqrt{\mu \cdot \varepsilon}}$$

The exponential decay of the medium content of each particle *core* by the *Propagated Outward Flow* of *medium* means that because of the gradual augmentation of the μ_0 and ε_0 the speed of light, c, is itself decaying, equation *(4-2)*.

(4-2)
$$c(t) = c_{[t=0]} \cdot \varepsilon^{-t/\tau}$$

The universal decay redshift occurs because we observe ancient light traveling at the speed at which it was originally emitted, a speed significantly larger than our present local speed of light because at the earlier time it was emitted it was less decayed. We observe the greater speed as longer wavelengths in the light. The formulation for the universal decay redshift, z_τ, of light that was emitted at time $t = T$ after the "Big Bang" and is observed at time $t = now = age\ of\ the\ Universe$ is as follows.

(4-3)
$$z_\tau = \frac{\lambda_T - \lambda_a}{\lambda_a} \quad T = \text{time since Big Bang that light observed was emitted}$$

$$= \frac{c(T) - c(a)}{c(a)} \quad a = \text{age of Universe} = \text{time now since Big Bang}$$

$$= \frac{c_{[t=0]} \cdot \varepsilon^{-T/\tau} - c_{[t=0]} \cdot \varepsilon^{-a/\tau}}{c_{[t=0]} \cdot \varepsilon^{-a/\tau}}$$

$$= \frac{\varepsilon^{-T/\tau}}{\varepsilon^{-a/\tau}} - 1$$

A number of years ago, in the late 20th Century, the estimates of astronomers and astrophysicists were that the earliest galaxies took about 2½ - 3 billion years to form, that is, that they did not appear until *2.5-3.0 billion years* after the Big Bang. Those estimates were based on analysis of the processes involved in star formation and in the aggregation and "clumping" of matter in the early universe.

Since then improved equipment and techniques [e.g. Keck and Hubble telescopes and gravitational lensing] have resulted in reports of observation of early galaxies having stars that formed as early as *300 million years* after the Big Bang according to the Hubble Law. Such a major reduction of earlier estimates to such a brief time and so soon after the Big Bang would appear to be questionable, but it is a result of the Hubble Law.

The Universal Exponential Decay completely resolves the problem of sufficient time after the Big Bang for the earliest stars to form. No matter how high the observed z may be the beginning of the exponential decay preceded it by the required star formation time. We may never know how long ago the Big Bang happened and we may never know the requisite time for the earliest stars to form, but we can know how long ago the oldest stars observed formed and we can know that the Big Bang took place a while before then.

Figure 4-1 below is a plot of the Universal Decay with the age of the Universe estimated to be *a = 30 gyrs*. That value is chosen to account for redshifts greater than *z = 11* having been observed and to allow for the possibility of still larger redshifts being observed.

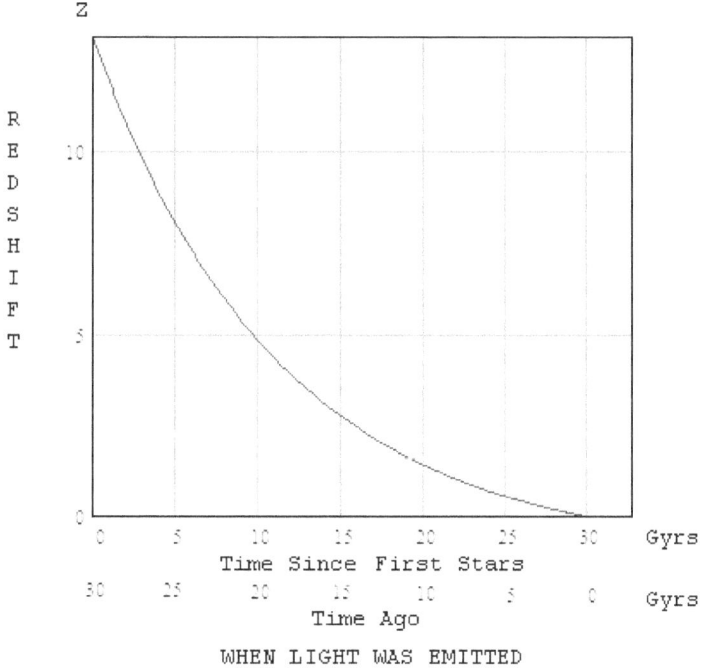

Figure 4-1
The Universal Exponential Decay

COMPARISON – HUBBLE VS. THE UNIVERSAL DECAY

Figure 4-2, below compares the Hubble Law, Figure 1-3 and the Universal Exponential Decay Figure 4-1 above.

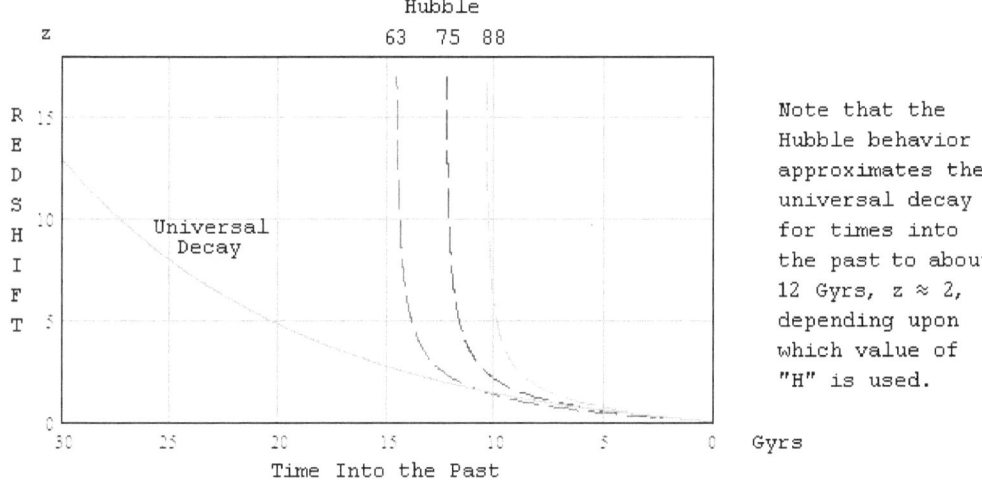

Note that the Hubble behavior approximates the universal decay for times into the past to about 12 Gyrs, z ≈ 2, depending upon which value of "H" is used.

Figure 4-2
Comparison: Hubble Law vs. Universal Exponential Decay

The above figure makes clear the reason for current cosmology estimates of the age of the universe being $13.8\ Gyrs$; it is the asymptotic behavior of the relativistic Hubble redshift formulation. Unfortunately, the currently more favored value for the Hubble "Constant", $H_0 = 73$ fails to correspond well to the currently favored age of the universe of $13.8\ Gyrs$.

The astral objects that we observe do exhibit a redshift caused by the Doppler Effect; however that is a minor part of the total redshift. The actual velocities involved are less than the speed of light. The primary cause of the observed redshifts is the Universal Exponential decay.

The next Part III presents four independent phenomena which
Evidence the Universal Exponential Decay:
Evidence Its Anomalous Acceleration a_A.

PART III -- THE PHYSICAL EVIDENCE FOR THE UNIVERSAL EXPONENTIAL DECAY

- *The Galactic Rotation Curves Anomaly*

- *The Pioneer Anomaly*

- *The Flyby Anomaly*

- *The Dark Flow Anomaly*

- *The Common Cause of the Anomalies – The Universal Decay*

- *Calculating The Anomalous Acceleration from The Decay Rate*

SECTION 5

Introduction to the Physical Evidence

THE ANOMALOUS ACCELERATION

There is throughout the Universe an acceleration that is not yet recognized let alone identified or understood although it has been (unknowingly) detected in that some of its effects have been observed and reported on. That acceleration [which is in addition to that of natural gravitation and which has no connection with gravitation] is small, centrally directed in every system that it appears, and acts independent of distance. The cause of that acceleration will be addressed shortly; however the purpose of the present evidence is to demonstrate that that anomalous acceleration exists and operates throughout the Universe.

THE SCIENTIFIC PROBLEM OF THE NEW AND UNKNOWN

The poet, Samuel Taylor Coleridge, in a philosophical lecture delivered in 1819, said, "Whoever is acquainted with the history of philosophy during the last two or three centuries cannot but admit that that there appears to have existed a sort of secret and tacit compact among the learned not to pass beyond a certain limit in speculative science. The privilege of free thought, so highly extolled, has at no time been held valid in actual practice except within this limit.

When new phenomena present a new challenge by calling for new, original hypotheses, whether because of the absence of any current hypothesis or because of the failure of all prior attempted hypotheses, the response of the science community at large, sometimes, is to ignore the new phenomena and the new data, to in effect dismiss them from the scientific search for knowledge.

Because it is an active creative process of new original material, the developing of new potentially valid hypotheses is more difficult than the observation of phenomena and the testing of resulting hypotheses, which is a relatively passive process.

Four independent unrelated phenomena, none of which has an established explanation nor a successful hypothesis, have now been extensively observed and a large amount of data substantiating the phenomena have been developed. However, with the exception of the search for "dark matter" interest in further investigating the phenomena seems to have disappeared as has the developing of badly needed potential hypotheses.

Even in the case of "dark matter" the current interest is solely in attempts to detect and identified that hypothesized substance and, in spite of that there now has been going on a century of that search with no positive results, there has been no interest in investigating hypotheses alternative to that of "dark matter".

The <u>four independent unrelated phenomena</u> can be shown to be the result of the same common cause operating on each to produce the diverse independent phenomena. The four phenomena are as follows.

- In 1933 F. Zwicky reported that the rotational balance of gravitational central attraction and rotational centripetal force in galaxies appeared to be out of balance, that a small additional centrally directed acceleration of unknown source appeared to be needed and to be acting. Numerous galactic rotation curves confirm that there is such an anomalous acceleration present and necessary in all rotating galaxies.

- In 1998 the Pioneer Anomaly was first reported. The anomaly is a small acceleration, centrally directed [toward the Sun], constant, distance independent, and of unknown cause, observed in the tracking of the Pioneer 10 and 11 spacecraft from launch until their near departure from the Solar System.

- In 2008 the Flyby Anomaly was first reported. The anomaly is unaccounted for changes in spacecraft speed, both increases and decreases, for six different spacecraft involved in Earth flyby from December 8, 1990 to August 2, 2005.

- Also in 2008 a previously unknown large scale flow of galaxy clusters all in the same direction toward "the edge" of the observable universe, the Dark Flow anomaly, was first reported. The mysterious motion, originally noted in 2008 using the three-year WMAP survey, is now [2010] confirmed by a more comprehensive five-year study .

- A fifth phenomenon and the earliest, astral Redshifts discovered by E. Hubble before the above, turns out to be primarily related to the same underlying phenomenon causing the others.

The same common cause operating on each is a locally centrally directed, small acceleration that is non-gravitational, distance independent, constant, and unaccounted for.

Next: The Galactic Rotation Curves Anomaly.

SECTION 6

The Galactic Rotation Curves Anomaly

THE ANOMALOUS ACCELERATION IN ALL ROTATING GALAXIES

In general, galaxies are rotating systems, a balance of gravitational attraction $[G \cdot M \cdot m / R^2]$ and centripetal force $[m \cdot V^2 / R]$ maintaining the structure. A curve or plot of such rotational velocity, V, versus path radius, R, is termed a Rotation Curve.

When the central mass is far greater than the orbiting masses the dynamics are such that the orbital velocities are inversely proportional to the square root of the radial distance from the center mass $[V = (G \cdot M / R)^{\frac{1}{2}}]$, as for example in our solar system and as illustrated in Figure 6-1, below. Such rotational dynamics and rotation curves are referred to as Keplerian.

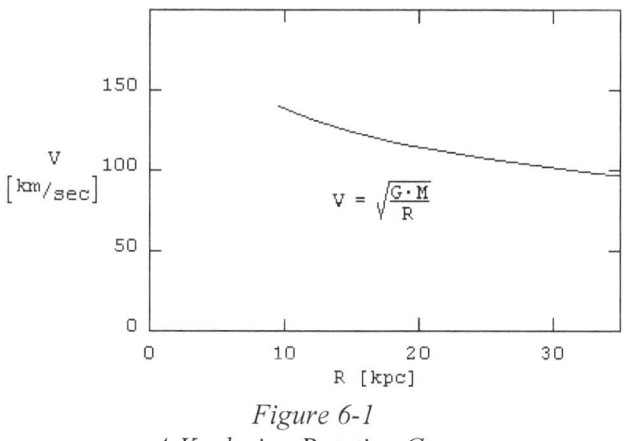

Figure 6-1
A Keplerian Rotation Curve

In the case of a solid sphere of uniform density, ρ, throughout, all parts must move at rotational velocities directly proportional to radius as illustrated in Figure 6-2, below.

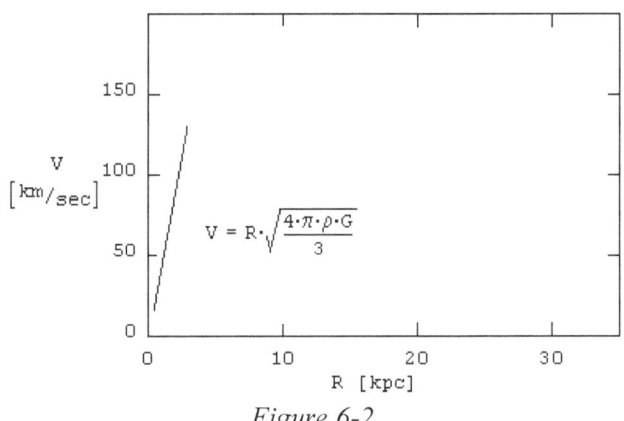

Figure 6-2
The Rotation Curve of a Solid Sphere of Uniform Density

The form of galaxies as we are able to directly observe them is that of a fairly spherical star-dense central core and a transition from that to the much more extensive flat disk of a far smaller density of more widely dispersed stars. The portion of galactic rotation curves that pertains to the dense central core of the galaxy would be expected to exhibit approximately the same velocity-proportional-to-radius form as illustrated for a solid sphere in Figure 6-2, above.

Likewise, the more dispersed flat disk, minor in mass compared to the dense central core, would be expected to exhibit the Keplerian form of Figure 6-1, above. The expected form of galactic rotation curves would be the two combined with a smooth transition between as Figure 6-3, below.

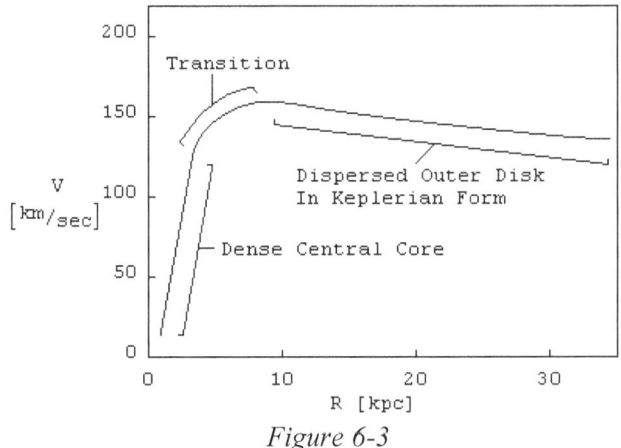

Figure 6-3
The Expected Form of Galactic Rotation Curves

For galaxies that present themselves in an edge view of the thin disk not as their spiral or globular spread in space, it is possible to measure the rotational velocities and obtain a rotation curve. We see one end of the presented flat disk moving toward us relative to the center and the other end moving away. The rotational velocities are measured along the galactic diameter represented by our view of the disk by observing the variations in redshift, those variations being a Doppler effect.

Galactic rotation curves so obtained do not exhibit the expected Keplerian form, an inverse square root of radius, in the region after the transition. Rather, they exhibit a flat form, that is, they there exhibit rotational velocity independent of radius. The overall curve, after the portion pertaining to the dense central core of the galaxy, is a transition to a flat curve in the region corresponding to the spread-out galactic disk as in Figure 6-4, below.

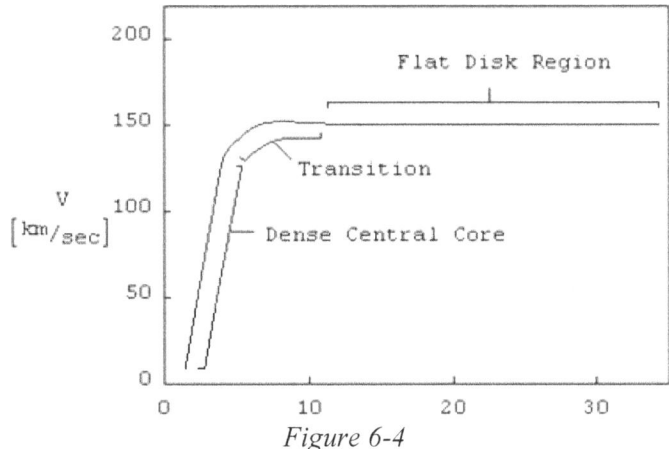

Figure 6-4
A Typical Galactic Rotation Curve as Observed

Because the form of the flat portion of galactic rotation curves lies between the case of a dominant central mass, as in the Keplerian inverse square root of radius form [Figure 6-1], and the case of a uniformly dense mass, with its direct proportion to radius form [Figure 6-2], it has been inferred that matter that we have not observed must be present similarly distributed within the galaxy.

That is, it is inferred that unobservable matter must be distributed in the galaxy in a manner that lies between the matter distribution of a dominant central mass [the Keplerian case] and that of a uniformly dense mass [the direct proportion to radius case] as a halo of "dark matter" which causes the rotation to take the form that the rotation curve exhibits. Thus arose the "dark matter" hypothesis.

No explanation has been offered for why the inferred "dark matter", while performing a gravitational function in the galaxy nevertheless fails to be distributed in the same manner as the "visible matter" in a fairly spherical dense central core with a transition from that to a much more extensive flat disk which has a far smaller density of more widely dispersed stars

What the rotation curves demonstrate is the existence of an <u>acceleration</u> that is not accounted for.

That acceleration can be identified as follows. A constant acceleration, $\Delta a_{Anomalous} = 8.4 \cdot 10^{-8}\ cm/sec^2$, expressed as Δa_A in the figure below, acting alone as a gravitational acceleration maintaining a mass in orbit, would produce a rotation curve of its own as in Figure 6-5, below. That acceleration, a_A, is directed toward the center of the galactic rotation just as is the accelerations of Figures 6-1 and 6-2.

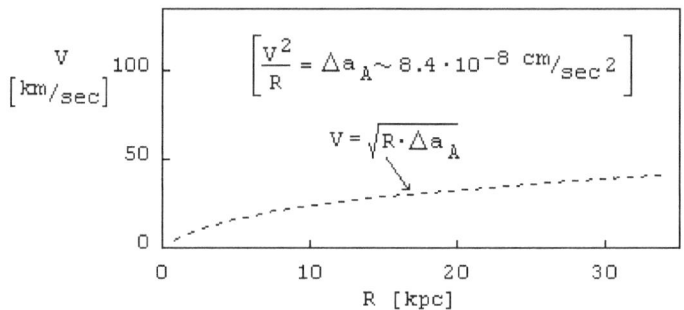

Figure 6-5
The Rotation Curve of $a_{Anomalous}$ Acting Alone

That rotation curve is of the correct form and magnitude to convert a galactic rotation curve exhibiting a Keplerian form [as in Figure 6-1] to a flat one [as in Figure 6-4]. That is, the rotation curve of a_A exhibits **V**elocity <u>directly</u> proportional to the square root of **R**adius and the Keplerian rotation curve exhibits V <u>inversely</u> proportional to the square root of R.

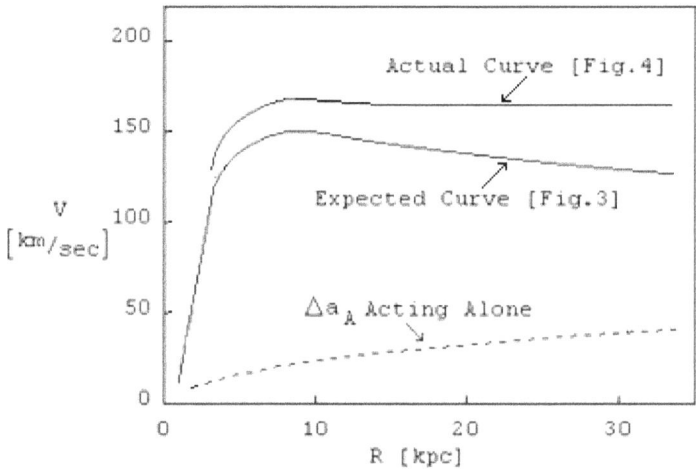

Figure 6-6
The Anomalous Acceleration, $a_{Anomalous}$, Acting Alone
Superimposed on the Expected and Actual Rotation Curves [Figures 6-3 & 6-4]

The two effects tend to cancel and leave a flat rotation curve. With the naturally occurring typical rotation curve modified by the addition of $a_{Anomalous}$ the rotation curve becomes flat, as illustrated in Figure 6-6, above, by superimposing the curves.

The rotation curves do not demonstrate the existence of a specific hypothesized cause such as dark matter; <u>they only demonstrate</u> the presence of a previously unknown acceleration, an acceleration unaccounted for by any known physical effects. It will be shown further below that the Universal Exponential Decay produces such an effect.

The hypothesized "dark matter" as the cause of the rotation curves behavior by supplying an additional gravitation is defective in that the dark matter is required to distribute itself differently than the other galactic matter in order to perform the function for which it was hypothesized.

After about a century of efforts to demonstrate the validity of the dark matter hypothesis by detecting identifiable dark matter, all without any success, an alternative cause of the galactic rotation curves behavior is called for.

This is the first appearance of a small acceleration (in addition to that of natural gravitation), centrally directed and independent of distance causing an anomaly

Next: The Pioneer Anomaly

SECTION 7

The Pioneer Anomaly

REVIEW OF THE PIONEER ANOMALY

The "anomalous acceleration" of the Pioneer 10 and 11 spacecraft was first reported in 1998. A weak, long-range acceleration toward the Sun had been observed in the Pioneer spacecraft. No satisfactory explanation has been found in spite of diligent efforts by a number of groups of scientists.

Pioneer 10 and 11 were launched in approximately opposite directions relative to the Sun and on paths that were to take them out of the Solar system. They were launched in 1972 and 1973 respectively and provided data into July 1998 and July 2000 respectively when they were far out at the extreme of the Solar System.

Each of them was spin stabilized with the spin axes running through the center of the dish antennae. That and their great distances from the Earth minimized the number of Earth-attitude reorientation maneuvers required, which enabled the precision of the acceleration data.

Conclusions reported with regard to the "Pioneer anomalous acceleration", notated as a_p, are as follows.

1. It is a real acceleration not pseudo.
2. It occurs in both Pioneer 10 and Pioneer 11 and is the same in both.
3. The researchers can find no mechanism or theory that explains it.
4. It is a line of sight constant acceleration of the spacecraft toward the Sun; that is, while always directed toward the Sun the magnitude of the acceleration, unlike the Sun's Newtonian gravitation, does not vary with distance from the Sun.
5. The measured magnitude of the acceleration was found to be per equation *(7-1)*.

(7-1) $a_p = (8.74 \pm 0.94) \times 10^{-8}$ $cm/s2$.

6. That result is in agreement, within its precision, with the anomalous acceleration $a_A = 8.4 \cdot 10^{-8}$ $cm/sec2$, for the galactic rotation curves of Figure 6-5.

7. Sources of systematic error:

- external to the spacecraft [e.g. solar wind, radiation]; and
- internal to the spacecraft [e.g. gas leakage]; and
- in the computational system [e.g. model accuracy, consistency]

have all been thoroughly addressed and are reflected in the equation *(7-1)* error range.

The Pioneer Anomalous Acceleration

The Pioneer Anomaly is a small acceleration of $(8.74 \pm 0.94) \times 10^{-8}$ $cm/sec2$, centrally directed [toward the Sun], constant, distance independent, and of unknown cause. The evidence for it is abundant tracking data that have been reviewed and re-reviewed in search of error with the result that the effect is highly validated.

If caused by the universal exponential decay the anomalous acceleration would be the same as the rate of change [decay] of the speed of light, *c*, as follows. The decay of *c* is as equation *(7-2)* and its rate of change is equation *(7-3)*.

(7-2) $\quad c(t) = c(0) \cdot \varepsilon^{-t/\tau}$

(7-3) $\quad \dfrac{d[c(t)]}{dt} = -\dfrac{c(0)}{\tau} \cdot \varepsilon^{-[t/\tau]}$

$\quad \dfrac{d[c(now)]}{dt} \equiv \dfrac{dc}{dt} = -\dfrac{c(now)}{\tau} \cdot \varepsilon^{-[0/\tau]} = -\dfrac{c(now)}{\tau} = -\dfrac{c}{\tau}$

Its value, using the value for τ of equation *(3-10)*, is equation *(7-4)*.

(7-4) $\quad \dfrac{dc}{dt} = -\dfrac{c}{\tau} = -\dfrac{2.99792458 \cdot 10^{10} \; cm/s}{3.57532 \cdot 10^{17} \; s} = -8.38505 \cdot 10^{-8} \; cm/s^2$

That result is in agreement, within their precision, with the reported anomalous acceleration inward of equation *(7-1)* and with the inward anomalous acceleration $a_A = 8.4 \cdot 10^{-8} \; cm/sec2$, for the galactic rotation curves of Figure 6-5. That validates that the effect is caused by the universal decay and validates the factor, *F*, and the time constant, τ of Section 3.

The only difference between the Pioneer Anomaly acceleration and the galactic rotation curve anomalous acceleration is that in the Pioneer case the acceleration is directed toward the Sun, the dominant factor in the mechanics of the Pioneer spacecrafts' motion whereas the galactic rotation curve anomalous acceleration is directed toward the rotational center of the galaxy, the dominant factor in the mechanics of galaxy rotation.

Next: The Flybys Anomaly

SECTION 8

The Flyby Anomaly

THE FLYBY ANOMALY

In March 2008 anomalous behavior in spacecraft flybys of Earth was reported in an article entitled "Anomalous Orbital-Energy Changes Observed during Spacecraft Flybys of Earth".

The data indicate unaccounted for changes in spacecraft speed, both increases and decreases, for six different spacecraft involved in Earth flybys from December 8, 1990 to August 2, 2005. These anomalous energy changes are a function of the incoming and outgoing geocentric latitudes of the asymptotic spacecraft velocity vectors and further indicate that a latitude symmetric flyby does not exhibit the anomalous speed change. The article states that, "All ... potential sources of systematic error .... [have been] modeled. None can account for the observed anomalies.... "Like the Pioneer anomaly ... the Earth flybys anomaly is a real effect .... Its source is unknown."

A phenomenon like that involved in galactic rotation curves and in the Pioneer Anomaly would account for the highly varied occurrences of the flyby anomaly: a small acceleration [in addition to that of natural gravitation], centrally directed and independent of distance; that is a modest and otherwise unknown acceleration directed toward the core center of the Earth, the principle body involved, the dominant factor in the mechanics of the flyby of Earth.

ANALYSIS OF THE FLYBY PARAMETERS

To observe the relation to the Flyby Anomaly of an otherwise unknown or un-detected anomalous, centrally directed, distance independent acceleration the first step is to consider a simple spacecraft pass of Earth where the pass is all at zero latitude as shown in Figure 8-1, on the following page. In the vectors analysis part of the figures A is the full anomalous acceleration, C is its component parallel to the direction of motion of the satellite, and θ is the angle between the direction of action of those two.

When the spacecraft is at a great distance out from Earth the spacecraft's motion is close to being directed toward the center of the Earth but not exactly so. A centrally directed acceleration there analyzed into components parallel and perpendicular to the spacecraft's motion would show most of the centrally directed anomalous acceleration acting to increase the spacecraft's speed.

a. <u>Polar View - Flyby</u>

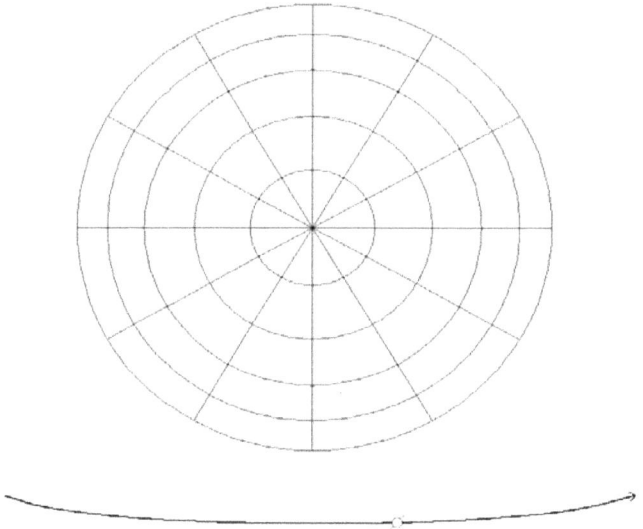

b. <u>Polar View - Anomalous Acceleration Vectors</u>

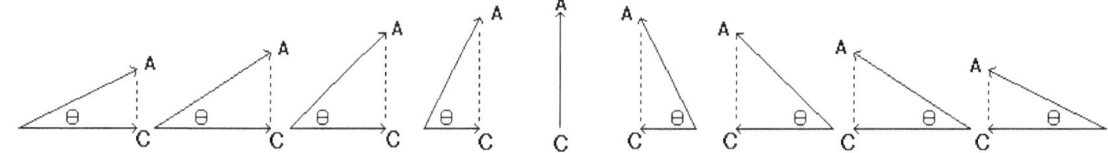

Here the acceleration phase and the deceleration phase are of equal magnitude and offset each other.

c. <u>Equatorial View</u>

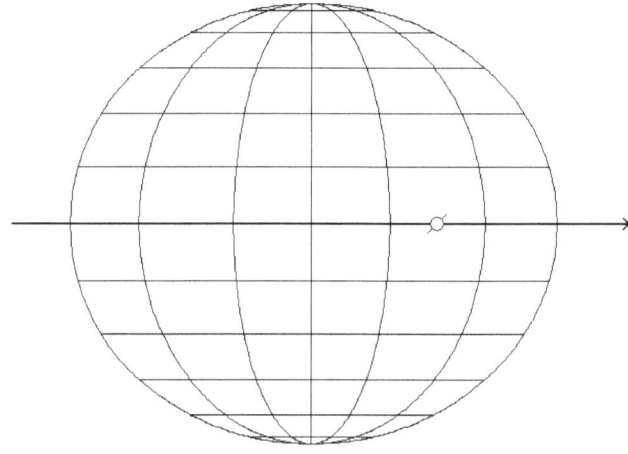

Figure 8-1
A Zero Latitude Pass

As the spacecraft travels nearer to Earth that component parallel to its motion decreases, becoming zero at the closest approach to Earth. From that point on the parallel component acts in the opposite direction on the spacecraft, that is its effect is to decelerate the spacecraft not accelerate it. Ultimately, in this example, the anomalous acceleration and anomalous deceleration experienced by the spacecraft become equal and cancel each other out leaving as the only flyby effect the gravitational boost, due to another effect, that is the overall purpose of the flyby.

Of the full centrally directed acceleration, A, the component, C, parallel to the path of the flyby in this case is

(8-1)   C = A·Cos[θ]

which is apparent if the flyby path is a straight line. However, the actual flyby path is somewhat curved by the Earth's gravitation. But, the anomalous acceleration is always centrally directed toward the core of the Earth so that C is nevertheless as stated.

Equation (8-1) is valid when the flyby pass is solely at zero latitude. However, if other than zero the latitude of the flyby pass has a significant effect on the magnitude of C, the component of the overall centrally directed acceleration parallel to the spacecraft flight path. As latitude increases the magnitude of C, decreases. That is most easily visualized by imagining the flyby over the geographic north pole at 90° north latitude. There the centrally directed acceleration toward the center of the Earth has no component parallel to the flight path.

Therefore, for flyby paths at other than zero latitude the effective value of A is A(λ) a function of latitude, λ, as equation (8-2)

(8-2)   A = A(λ) = A·Cos[λ]

so that equation (8-1) then becomes equation (8-3) the full expression for the extent to which the centrally directed anomalous acceleration actually accelerates or decelerates the spacecraft.

(8-3)   C = A·Cos[λ]·Cos[θ]

The gross effect of latitude can be evaluated by examining three cases:

A - The flyby path is symmetrical relative to the equator so that the latitude effect in the first half of the flyby, θ = 0° to 90°, is exactly offset or balanced by the second half of the flyby, θ = 90° to 180°. This case is essentially the same as presented in Figure 8-1, above.

B - The flyby path starts at low latitude and finishes at high latitude, Figure 8-2 on the following page.

C - The flyby path starts at high latitude and finishes at low latitude, Figure 8-3 on the second following page.

Per the equations and Figure 8-1 in the first half of the flight path the effect of the anomalous, centrally directed acceleration is to increase the speed of the spacecraft whereas the effect in the second half of the flight path is to decrease the spacecraft's speed. By its definition Case A produces no net anomalous acceleration or deceleration of the spacecraft because the first and second halves of the flight path balance and offset each other.

a. <u>Equatorial View - Flyby</u>

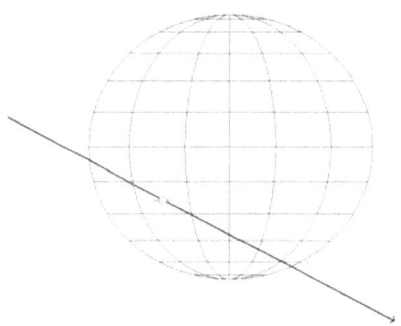

b. <u>Equatorial View - Flyby, Rotated</u>

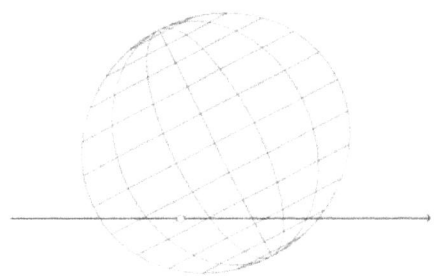

c. <u>Anomalous Acceleration Vectors</u>

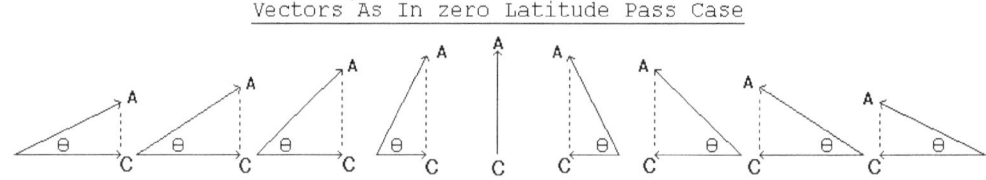

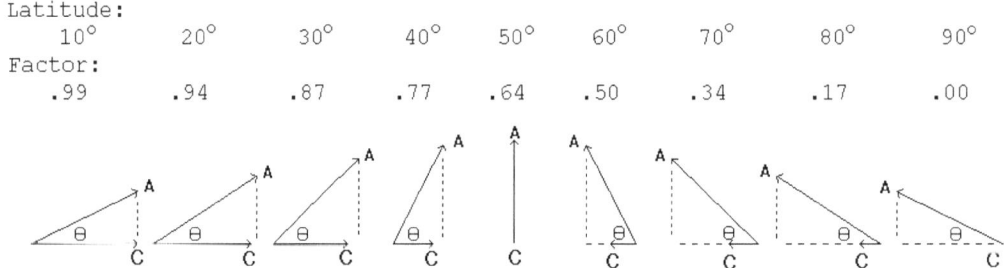

The result n this case is a net acceleration
[to the right in the diagrams].

Figure 8-2
A Pass at Increasing Latitude

a. <u>Equatorial View - Flyby</u>

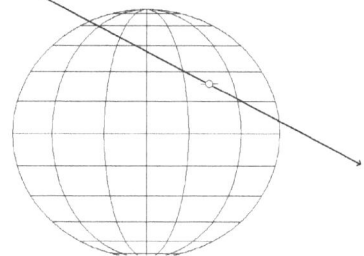

b. <u>Equatorial View - Flyby, Rotated</u>

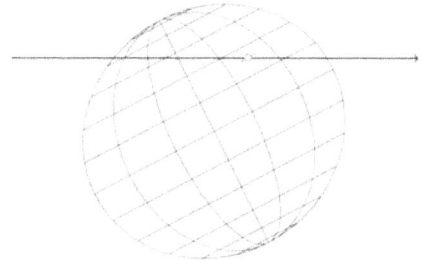

c. <u>Anomalous Acceleration Vectors</u>

Vectors As In zero Latitude Pass Case

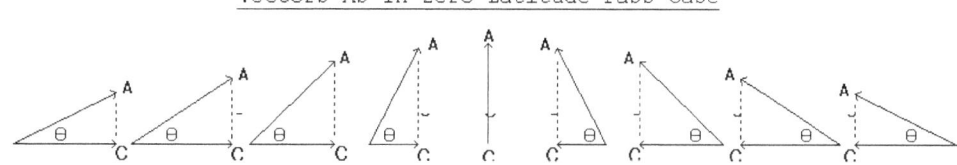

Above Vectors as Further Reduced by Non-Zero Latitude
Reduction Factor = Cosine[Latitude]

Latitude:	90°	80°	70°	60°	50°	40°	30°	20°	10°
Factor:	.00	.17	.34	.50	.64	.77	.87	.94	.99

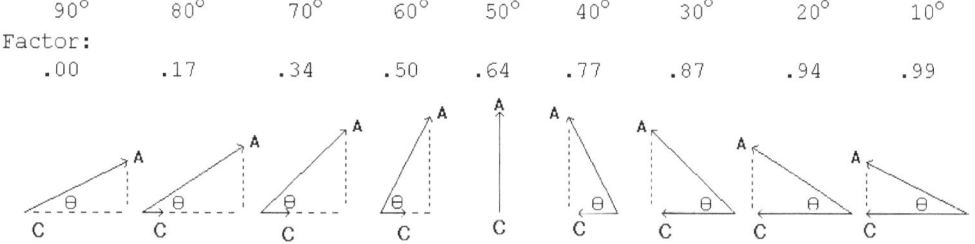

The result in this case is a net deceleration, that is an acceleration
to the left in the diagrams, against the direction of the flyby.

Figure 8-3
A Pass at Decreasing Latitude

In Case B, the first half, i.e. the acceleration half, of the flight path is at low latitude where the latitude effect only modestly reduces the anomalous acceleration magnitude. But for that case and path the second half, i.e. the deceleration half, of the flight path is at a high latitude where the latitude effect greatly reduces the anomalous acceleration magnitude. The net effect is a relatively large acceleration followed by a lesser deceleration for a net increase in the spacecraft's speed.

In Case C, the effect is just the reverse of that in Case B; the first, i.e. the acceleration, half of the flight path is at high latitude where the effect of the latitude greatly reduces the anomalous acceleration magnitude. But for that case and path the second, i.e. the deceleration, half of the flight path is at a low latitude where the effect of the latitude only modestly reduces the anomalous acceleration magnitude. The net effect is a relatively small acceleration followed by a greater deceleration for a net decrease in the spacecraft's speed.

Therefore, depending on the specific flight path of the spacecraft's flyby pass of Earth the spacecraft may experience an overall net anomalous acceleration or a net anomalous deceleration, those in various amounts depending on the specific encounter and the latitudes involved, and zero net modification if the path is perfectly latitude symmetrical.

Thus there is a small acceleration [in addition to that of natural gravitation], centrally directed and independent of distance producing the flyby anomaly; that is, a modest and otherwise unknown acceleration directed toward the core center of the Earth, the principle body involved, the dominant factor in the mechanics of the flyby.

And, that is another appearance of the same common small acceleration (in addition to that of natural gravitation), centrally directed and independent of distance appearing [here planet-wise in the Flyby Anomaly] and previously [solar system sun-wise in the Pioneer Anomaly] and again [galaxy-wise in the rotation curves anomaly].

Next: The Dark Flow Anomaly

SECTION 9

The Dark Flow Anomaly

THE UNIVERSE CONTEXT

Having found that there are small, centrally directed, distance independent, non-gravitational, same, anomalous accelerations appearing as a near Earth effect [the Flybys Anomaly], a Solar effect [the Pioneer Anomaly], and a galactic effect [galactic rotation curves], it can be expected that the same effect appear relative to every planet [and every planet's moons], every sun [star], every galaxy and every group of galaxies.

And, such a small, centrally directed, distance independent, non-gravitational, same, anomalous acceleration could be expected to appear for the universe overall, appear directed toward the center of the universe, the location of the origin, where it all began.

The universe began with the "Big Bang", an immense explosion radially outward in all directions, largely spherically symmetrically, from an original source "singularity".

We, residing on planet Earth, of star Sol, in one of several branches of spiral galaxy Milky Way, are located off some significant distance in "our general direction" from and relative to the location of that original singularity.

We can "see" or detect a large number of neighbor galaxies, distant and near, whose components, as did ours, proceeded outward from that "Big Bang" in directions slightly or significantly other than our particular direction.

But, there is a further mass of stellar bodies that proceeded outward from the "Big Bang" in directions away from us. What we can detect is only well less than half the total product of the "Big Bang".

The original location of the singularity, the origin, lies essentially at the center of the largely spherical volume of the source's product, the expanding universe. And the universe that we "see" lies largely to one side of that origin's location

THE TROUBLE WITH THE HUBBLE LAW

THE "DARK FLOW" ANOMALY

To the above list of three effects caused by a small, centrally directed, distance independent, non-gravitational, common, anomalous acceleration can now be added the "Dark Flow" as originally reported in 2008 and further analyzed in terms of extensive new data as reported in a NASA Goddard Release.

"Distant galaxy clusters mysteriously stream at a million miles per hour towards a single point in the sky, separate from the expansion of the universe, along a path roughly centered on the southern constellations Centaurus and Hydra." The phenomenon is dubbed by the researchers the "dark flow"."

"The clusters appear to be moving along a line extending from our solar system toward Centaurus / Hydra away from Earth. The distribution of matter in the observed universe cannot account for it. Its existence suggests that some structure beyond the visible universe -- outside our "horizon" -- is pulling on matter in our vicinity."

"Dark Flow" Galaxy Clusters and Flow Direction by Distance

- Clusters from 0.8 – 1.2 billion light-years away (250 to 370 megaparsecs)
- Clusters from 1.2 – 1.7 billion light-years away (370 to 540 megaparsecs)
- Clusters from 1.3 – 2.1 billion light-years away (380 to 650 megaparsecs)
- Clusters from 1.3 – 2.5 billion light-years away (380 to 755 megaparsecs)

The colored dots are galaxy clusters within one of four distance ranges. Colored ellipses show the direction of bulk motion for the galaxy clusters of the corresponding color.

Figure 9-1
The "Dark Flow"

This is indication of the overall universe's experiencing an anomalous centrally directed acceleration accelerating all the matter of the universe gradually back toward the location of its origin [as described above]. This "Dark Flow" is part of that centrally directed acceleration toward the location of the origin of the universe, a location at or just beyond the "edge" of the universe "see-able" by us.

A "map" of the universe that we "see" would look somewhat as in Figure 9-1, above, where the regions of galaxies studied involved in the "dark flow" are indicated in the large colored areas in red, yellow, green and blue.

Thus we have a small, centrally directed, distance independent, non-gravitational, common, anomalous acceleration acting directed at the center of every object system in the Universe: planets, stars, galaxies, and the overall Universe.

There Is a Common Cause of All These Varied Phenomena:
– *The Universal Exponential Decay* –
As Follows

SECTION 10

The Common Cause – The Universal Exponential Decay

A GENERAL ANOMALOUS ACCELERATION THROUGHOUT THE UNIVERSE

We have now found that there are small, centrally directed, distance independent, non-gravitational, same, anomalous accelerations appearing as a near Earth effect [the Flybys Anomaly], a Solar effect [the Pioneer Anomaly], a galactic effect [galactic rotation curves], and a Universe effect [the Dark Flow]. It can only be concluded that the same effect must appear relative to every planet [and every planet's moons], every sun [star], every galaxy and group of galaxies, and the universe overall. In other words as a general cosmic effect.

What could produce such a phenomenon ? What would cause there to be a Universe-wide occurrence of such same inward accelerations ?

Taken together, planet relative, star relative, galaxy relative, Universe relative, they collectively are a systematic contraction, a gradual reduction in the length component of every physical quantity in the universe. A general universal decay, a universal contracting which is the result of the *Propagated Outward Flow* gradually depleting the original supply of *medium* in the *core* of each particle of the Universe per Section 3.

In material reality such decays are exponential. There are myriad examples of such, for example: radioactive decay, the decay of electrical transients in circuits involving inductance, capacitance and resistance, the decay of motion transients in mechanical systems involving mass, spring and damping, the amplitude decay in a rung bell or a plucked string, etc. It is not unreasonable that a universe that began with an explosive "bang" follow that with a gradual exponential decay.

The Universal Exponential Decay is an exponential decay of the length dimensional aspect of all quantities in the universe. It involves the fundamental constants (c, q, G, h, etc.) and decay of any of those must be dimensionally consistent with the decay of the others. The dimension that is decaying is length, the $[L]$ dimension in the dimensions of, for example: the Planck constant, h, $[M \cdot L^2/T]$; the speed of light, c, L/T; and the gravitational constant, G, $[L^3/M \cdot T^2]$. The time constant of the decay is about $\tau = 3.57532 \cdot 10^{17}$ sec ($\approx 11.3373 \cdot 10^9$ years) per Section 3.

Objections that such an effect would conflict with the solar system's known planetary system performance per the highly accurate planetary ephemeris are a mistaken interpretation of the situation. Consider a planet in circular orbit around a sun as in Figure 10-1, below.

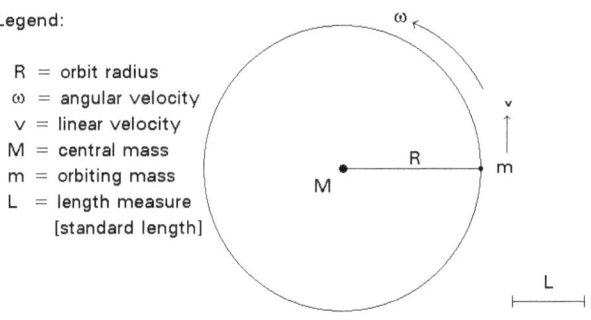

Figure 10-1

The relationship governing the motion is, of course, equation *(10-1)*, below.

(10-1) Centripetal Acceleration = Gravitational Attraction
Required   Acceleration
V^2/R (or) $R \cdot \omega^2$ = $G \cdot M/R^2$

Now, let the length dimensional aspect of all quantities decay, becoming gradually smaller with time [with the dimensions of all quantities expressed in the fundamental dimensions of mechanics, $[L]$, $[M]$, and $[T]$]. That is, let all lengths, $[L]$, decrease by being multiplied by the decay function, $D(t)$, per equation *(10-2)*, below. [For the present purpose the form of the decay function is irrelevant except that it must be a function of time. The decaying exponential is used because it is common in nature and is a complicated case.]

(10-2) $D(t) \equiv \varepsilon^{-[t/\tau]}$, where τ is the time constant of the decay

Then the quantities involved in equation *(10-1)* all change to as follows.

(10-3) <u>The Orbital Radius</u>, R, [dimension = L]

R becomes $R(t) = R(t=0) \cdot D(t)$

<u>The Gravitational Constant</u> [dimensions = $L^3/M \cdot T^2$]

G becomes $G(t) = G(t=0) \cdot [D(t)]^3$

<u>The Centripetal Acceleration Required</u> [dimensions = L/T^2]

$R \cdot \omega^2$ becomes $R(t) \cdot \omega^2 = [R(t=0) \cdot \omega^2] \cdot D(t)$

etc. The overall net effect is: R decays, the required centripetal acceleration decays in proportion, the gravitational attraction likewise decays in proportion, and ω is unchanged. Because the time constant and the start of all of the decays are identical and arise from the same Big Bang beginning, the physical laws inter-relationships remain unchanged and coordinated with each other.

Furthermore, we observers, using our measuring standard ruler, length L of the above Figure 10-1, would never detect any of the decay because our standard length would also be decaying at exactly the same rate, in the same proportion.

The point of this obvious mathematics / physics exercise is that a universal decay of the length aspect of all material reality would not conflict with the planetary ephemeris and would not even be detectable at all except in unusual circumstances such as the Pioneer and Flyby anomalies and the evidence of galactic rotation curves; nor would it interfere with the relative values of the fundamental constants and their interactions in physical laws.

Returning to the orbiting body of Figure 10-1, reproduced as Figure 10-2, below, the figure's annotations slightly modified, the development of the anomalous acceleration is very direct.

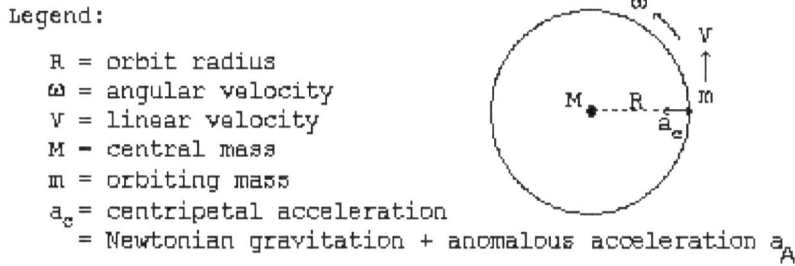

Legend:

    R = orbit radius
    ω = angular velocity
    V = linear velocity
    M = central mass
    m = orbiting mass
    a_c = centripetal acceleration
        = Newtonian gravitation + anomalous acceleration a_A

Figure 10-2

The Newtonian gravitation component of the centripetal acceleration, a_c, is only sufficient to maintain the orbit, to keep R constant, to prevent its increasing. For the orbiting body, m, to gradually approach the central mass, M, that is for R to decrease, additional inward acceleration is required.

That inward acceleration is the anomalous acceleration appearing as a near Earth effect [the Flybys Anomaly], a Solar effect [the Pioneer Anomaly], and a galactic effect [galactic rotation curves]. It is an unavoidable concomitant effect of the contraction of the length dimension *[L]* of R in the above example and of the systematic contraction, the gradual reduction in the length component, of every physical quantity in the Universe, of all material reality by the Universal exponential decay.

Then, what is the magnitude of that anomalous acceleration ?

Next: Calculating The Anomalous Acceleration from The Universal Exponential Decay's Decay Constant

SECTION 11

Calculating The Anomalous Acceleration from The Decay Constant

The Universal Exponential Decay is the gradual depletion of the *medium* supply in the *core* of every particle throughout the Universe, depletion by its *Propagated Outward Flow* of some of that *medium*. The flow at a decaying speed and wavelength results in all of the various decay effects, but the essential basic decay is the *core* content of *medium* decaying by the gradually decaying outward flow of *medium* from it at a decaying speed.

The speed of the flow is what we call the speed of light. It is not that the speed of traveling light gradually decays. Rather it is that each instant of light carrying medium as successively emitted from its source core is successively at a slightly further decayed speed than its predecessor.

That decaying speed of the flow is per equation *(11-1)*.

(11-1) $\quad c(t) = c(0) \cdot \varepsilon^{-t/\tau} \quad$ a velocity.

Its rate of change, that is its acceleration [in this case deceleration], is its first derivative, as equation *(11-2)*.

(11-2)
$$\frac{d[c(t)]}{dt} = -\frac{c(0)}{\tau} \cdot \varepsilon^{-[t/\tau]} \quad \text{in general.}$$

$$\frac{d[c(now)]}{dt} \equiv \frac{dc}{dt} = -\frac{c(now)}{\tau} \cdot \varepsilon^{-[0/\tau]} = -\frac{c(now)}{\tau} = -\frac{c}{\tau} \quad \text{at time now}$$

$$a_A = -\frac{c}{\tau} = -\frac{2.99792458 \cdot 10^8}{3.57532 \cdot 10^{17}} \quad \text{now}$$

$$= -8.38504 \cdot 10^{-10} \; \text{m}/\text{s}^2 \quad \text{the anomalous acceleration}$$

That result is that the rate of change, the acceleration of the Universal Decay, is an inward acceleration, the universal contraction, that matches the experimentally measured anomalous acceleration inward, a_P, of $(8.74 \pm 0.94) \times 10^{-10}$ m/s2 of the Pioneer Anomaly of Section 7 within the limits of its (just above) precision.

It is the rate of the universal contraction, the un-accounted for centrally directed acceleration demonstrated in galactic rotation curves, the Flybys Anomaly and the Dark Flow Anomaly.

Because the decay time constant is so large relative to human life times the decay appears to us to be a constant unchanging state.

Because everything including our instrumentation, our measurement standards, our atoms and ourselves are all experiencing the same decay, the decay is unnoticeable to us and is generally undetectable by us except for unusual circumstances such as the anomalies presented above.

CONCLUSION

The calculation of the anomalous acceleration, a_A, is dependent on the value of the decay time constant, τ, derived in Section 3, equations *(3-6)*, *(3-9)*, and *(3-10)*. That it gives a resulting value that matches the experimentally observed value of the anomalous acceleration obtained from the Pioneer Anomaly and galactic rotation curves is confirmation of the Universal Exponential Decay and the value of its time constant, τ.

Next: Further Experimentally Validating the Universal Exponential Decay

PART IV -- CONCLUSION

- Further Validating the Universal Decay

- Conclusion

- The Cosmos' Expansion From The Origin To The Present

- References

SECTION 12

Further Experimentally Validating the Universal Decay

Validating The Universal Decay

As developed in Section 4, each particle's *Propagated Outward Flow* contains, carries within it, has its speed of propagation determined by, its own μ_0 and ε_0. Traveling *medium*, and therefore the light being carried by it, do not further decay once they are emitted from their source particle; rather, their speed is permanently set by their then embedded own μ_0 and ε_0. Because of the decay process at the emitting particle its following emissions are at successfully slower speeds.

Because the observed speed of light is decaying, light emitted long ago is faster than our present, local contemporarily emitted light, which causes the ancient light to appear to us to have a longer wavelength, that is, to be redshifted. [A small portion of redshifts, but not more than a minor portion, is due to the Doppler Effect of the astral sources' outward velocities, velocities always smaller than c.]

Aside from observation of redshifts, <u>each observation of which is actually an observation of the Universal decay</u>, there are two other specific experimental observations that can be conducted to verify the Universal Decay and the value of its decay time constant.

> 1- It can be tested that the speed of the light <u>from far distant astral sources</u> is larger than our contemporary local light speed. The earlier procedure of Michaelson and Moreley is now superseded by the modern procedure, which is to modulate the light beam placing brief markers along its path and use those markers to measure the time required for the light to traverse a known distance.

For that purpose the decay of the speed of light is as per equation *(12-1)*.

(12-1)
$$c(t) = c_{[t=0]} \cdot \varepsilon^{-\left[\frac{t}{\tau}\right]}$$

Similarly the value of the Planck Constant in traveling *medium*, traveling light, does not further decay from its value in the *medium*, light, at the time it was emitted from the source particle; rather, its value is permanently there, then embedded.

2- It can be tested that the Planck Constant of the light <u>from distant astral sources</u> is larger than our contemporary Planck Constant, h, using the photoelectric effect. Measuring the retarding potential that reduces the photoelectric current to zero, for light spectrally selected of a specific frequency, plots [for a set of different frequencies] as diagonal straight lines whose slope is the Planck Constant of that light as in Figure 12-1.

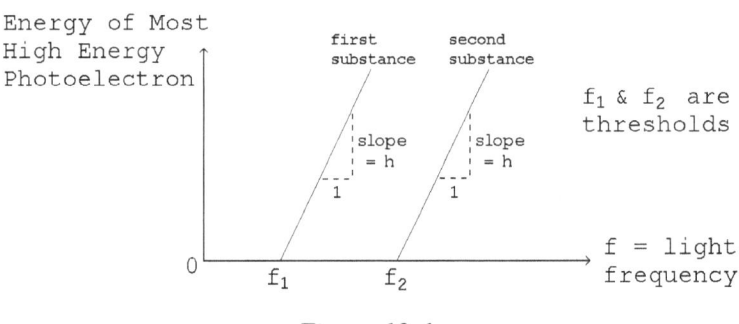

Figure 12-1

The dimensions of the Planck Constant, h, are $[M \cdot L^2/T]$ as compared to the dimensions of light, $[L/T]$. Therefore, whereas the decay of light is as equation *(12-1)*, above, the decay of the Planck Constant is per equation *(12-2)*.

(12-2)
$$h(t) = h_{[t=0]} \cdot \varepsilon^{-[t/\tau]} \cdot \varepsilon^{-[t/\tau]} = h_{[t=0]} \cdot \varepsilon^{-[2 \cdot t/\tau]}$$

The behavior of the decay of light and the decay of the Planck Constant are as in Figure 12-2 in which the scales are logarithmic, not linear.

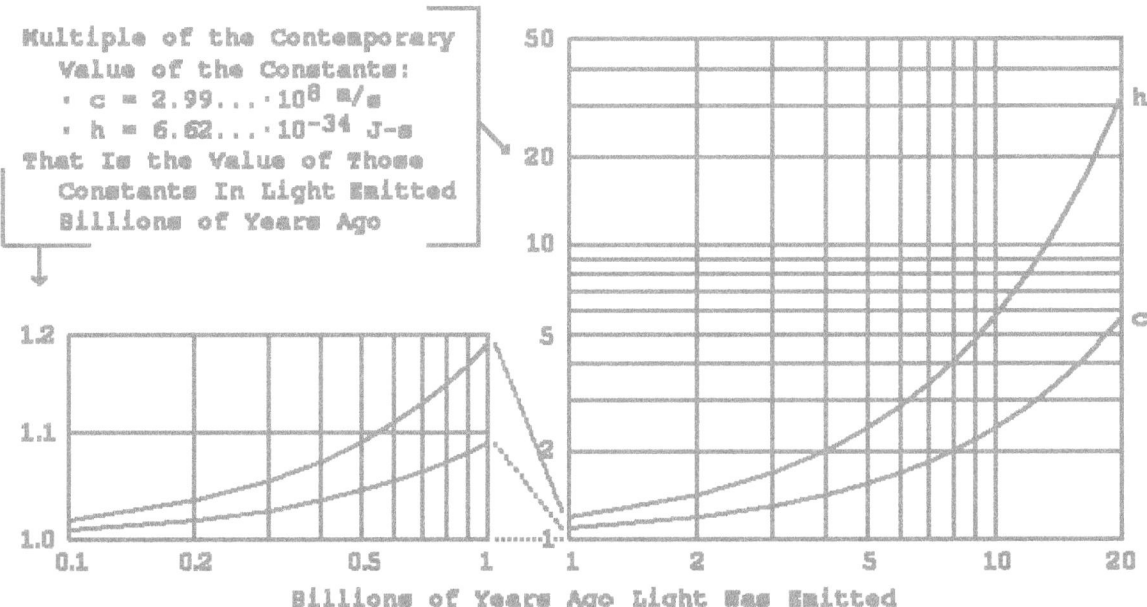

Figure 12-2

SECTION 13

Conclusion

The Universal Exponential Decay not only works better than the Hubble theory, it is also much more well validated by other phenomena than is the Hubble theory.

WHY THE HUBBLE THEORY MUST BE ABANDONED

The Hubble theory problems include whether its constant, H_0, is truly a constant at all, which goes beyond the existing problem of finding a value for it. Other problems are the asymptotic nature of the function, which severely distorts the interpreted age of the Universe, and that the underlying assumptions are unreliable. The Hubble theory requires accepting that early stars and galaxies formed at an extremely short time [cosmically speaking] after the Big Bang.

The biggest problem stems from that at the time the theory was developed values for z greater that `1.0` were inconceivable. Consequently the problem of the theory calling for recession velocities exceeding the speed of light did not arise until later upon the observing of just such values for z. By then the theory had been so well accepted and embedded in the practice of astronomy that the thought of abandoning it with no alternative on the horizon could not be seriously entertained.

The result was the resorting to the concept that it was expansion of space itself that was violating the speed limit set by light, and that space was doing that "legally". However, whether by "physical" motion or by being carried along by expansion of space the Hubble theory calls for actual separation distances from us to be increasing at a rate that is impossible.

WHY THE UNIVERSAL DECAY THEORY MUST BE ADOPTED

The Universal Decay analysis of redshifts has none of the problems of the Hubble theory: its only constants are the well established speed of light, c, and the natural logarithmic base, ε; its age of the Universe and time for the earliest stars and galaxies to form are reasonable; it relies on no unreliable assumptions and it calls for full respect for the speed limit of light speed. It readily accommodates the earliest stars and galaxies not forming earlier than `2-3 billion years` after the Big Bang.

[The Doppler Effect still causes a minor part of redshifts because the recession velocities must have that effect; however, those shifts are always due to recession velocities less than the speed of light and the dominant part of redshifts is that due to the Universal Exponential Decay].

THE TROUBLE WITH THE HUBBLE LAW

The Universal Decay is supported by the combined evidence of four anomalies none of which has a satisfactory resolution other than the Universal Decay: the Rotation Curves anomaly, the Pioneer anomaly, the Flyby anomaly, and the Dark Flow anomaly.

The Universal Decay provides a perfect cosmic "yardstick" for all distances to all cosmic objects, near and far, in that any objects' light observed here now carries the information as to its place on the exponential decay curve, the time it was emitted, in its specific speed and wavelength shift.

For What This Means
for the "Shape" of Today's Cosmos
see the following Section 14 -

The Cosmos' Expansion From The Origin To The Present

SECTION 14

The Cosmos' Expansion From The Origin To The Present

This analysis is of the mechanics of the travel of matter outward from its "Big Bang" source [some of it ultimately being we the observers] and of the mechanics of the travel of light from such material sources wherever they are at the time that the light that we later observe is emitted. Because the observing of light from very distant astral sources is the observing of light emitted billions of years ago, this analysis includes the question of how far back into the past it is theoretically possible to observe and of the ultimate fate of the universe.

1 - THE TRAVEL OF MATTER AND LIGHT

The first step is to develop formulations that describe the travel of the two different traveling entities, light and matter, at various times in the past from at the beginning to the present.

The travel of matter originated at the location of the "Big Bang" singularity and was initially radially outward from that location. While mass cannot travel at light speed the initial speed of the "Big Bang" product particles was sufficiently near the then [initial un-decayed] light speed so as to be taken as such as is developed below. Two effects then proceeded to slow the outward velocities: the decay of the speed of light [the upper limit on particle velocity] and the gravitational slowing [the centrally directed gravitational acceleration, caused by the total mass, decelerating the outward velocities].

The treatment here is of the estimated "average" or "typical" cosmic body [e.g. galaxy], treated as that from its initial form as myriad fundamental particles at the instant of the "Big Bang" -- the particles ultimately destined to form that particular "typical" body, through its form as we know it now. [While not of concern in the present analysis, once the outward travel began the particles experienced local gravitational effects in addition to the overall general slowing -- effects that deflected paths from being purely radially outward and that lead to "clumping" and the formation of structure in the universe.]

The travel of light originated from the above traveling matter, at its various locations and times throughout the universe from the first instant on. It was radially outward from wherever its source was at the time of emission. Its speed was the speed of light at the decayed value for the time after the "Big Bang" that the light was emitted.

a. *The Travel of Light Outward From Astral Sources*

Astral / cosmic source light emitted long ago was emitted at a higher "light speed" than our local contemporary light and continues to travel at that faster speed forever as explained earlier in section 15, beginning on page 255 at "ε, μ, and the Speed of Propagation". [Briefly, the decay is in the generation, in the source, not the propagation. That is the case because the emitted light carries within it its own propagation-determining permeability and dielectric constant, μ and ε. How could it be otherwise since light propagating outward into unoccupied space, into pure nothing, would certainly find no μ and ε there: in nothing?

On the other hand, the matter originating with the "Big Bang" cannot have traveled at light speed [because its mass would then be infinite] other than nearly so initially before being slowed by gravitation. Therefore, all cosmic source light has been traveling at greater speeds than the cosmic bodies that are home for observers of the light.

Consequently, the most ancient light that it would be theoretically possible for us to observe would be light from a cosmic source that exited the "Big Bang" in the diametrically opposite direction to that of the planetary home [or its components before they became the home planet] of we, the observers. That way, the ancient light has to travel a maximally greater distance from its location where and when emitted to our location where and when we observe it than did our planetary home have to travel from its location when the light was emitted to its location when we observe the light. In other words, ancient light is light that has been traveling a long time and, therefore, has traveled a great distance. The home of we, the observers cannot travel so fast and must, therefore, have a "head start" of distance to be able to arrive at the meeting place of light and observer at the same moment as the faster light. The largest "head start" is the handicapping of the light by placing its source diametrically opposite the location of the observers.

Standard International [SI] units are used; however the great range of magnitudes of the quantities considered calls for their being expressed sometimes in alternative astronomical units: time in `Gyrs = Years·10`^{9} rather than `seconds` and distances in `"our" G-Lt-Yrs = 10`^{9}`× [Light Years at our contemporary speed of light]` rather than `meters`. [Note: `G-Lt-Yrs` is always "our" `G-Lt-Yrs`.] Those are obtained by the following factors.

(14-1)
$$k_{time} = 60 \cdot 60 \cdot 24 \cdot 365 \cdot 10^9 \quad [sec/_{Gyr}]$$
$$k_{dist} = k_{time} \cdot [\text{"Our" Light Speed}]$$
$$= k_{time} \cdot [2{,}997{,}924.58 \cdot 10^8] \quad [meters/_{G\text{-}Lt\text{-}Yr}]$$

The actual practical precision of the calculations is limited to one or two significant figures by the nature of the estimates of quantities such as the density and mass of the universe and the universe's expansion velocities and their distribution. The greater number of significant figures indicated in the data tabulations do not signify greater precision of results. Rather, they are included because they are the actual data used in the calculations before rounding to the real precision, and they are the data used in generating the various graphs. They make the results presented completely reproducible.

For the present the age of the universe is taken to be unknown so that *Age* is a variable. Then the original speed of light, *c(0)*, at the instant of the "Big Bang", just before the first moment of the Universal Decay, is obtained using equation 14-2, the Universal Decay time constant, τ, [from earlier equation 3-9], in equation 14-3, the calculations, below.

(14-2)

$$\tau = 3.57532 \cdot 10^{17} \text{ sec} \quad [\approx 11.3373 \cdot 10^9 \text{ years}]$$

(14-3)

$$c(t) = c(0) \cdot \varepsilon^{-t/\tau} \text{ meters/sec} \quad \text{[light universal decay]}$$

$$c(Age) \equiv 2.997,924,58 \cdot 10^8 \text{ m/sec} \quad \text{["our" c, now]}$$

$$c(Age) = c(0) \cdot \varepsilon^{-Age/\tau} \quad \text{[set t = Age in first line]}$$

$$c(0) = c(Age) \cdot \varepsilon^{+Age/\tau} \quad \text{[solve for c(0)]}$$

$$= 2.997,924,58 \cdot 10^8 \cdot \varepsilon^{Age/\tau} \text{ m/sec}$$

Then, the speed of light at any arbitrary time, *t*, after the "Big Bang" for any arbitrary age of the universe, *Age*, is as follows.

(14-4)

$$c(t, Age) = c(0) \cdot \varepsilon^{-t/\tau}$$

$$= [2.997,924,58 \cdot 10^8 \cdot \varepsilon^{Age/\tau}] \cdot \varepsilon^{-t/\tau} \text{ meters/sec}$$

b. *The Travel of Cosmic Bodies Outward From the Origin of the "Big Bang"*

To determine the travel of cosmic bodies outward from the "Big Bang" one needs to know the initial velocities and the manner in which they subsequently were reduced by gravitation and other effects. The initial radially outward velocities were so close to the then speed of light as to be that speed for the practical precision here being used. That determination develops as follows.

(1) The Initial Radially Outward Velocities

The universe has existed for billions of years and is still expanding. Therefore, the initial velocity / energy of the "Big Bang" product particles must have been near, if not at or greater than, the escape velocity / energy. The escape velocity / energy for any one particle of the initial "Big Bang" universe is calculated as follows. [The calculation is done non-relativistically here and consequently produces apparent velocities much greater than that of light. They represent velocities nearly at light speed with greatly increased mass.]

Gravitational escape velocity is that velocity the kinetic energy of which just equals in magnitude the potential energy of position in the gravitational field for which the escape velocity is being determined. The non-relativistic escape velocity of a particle develops as follows.

(14-5)

$$\text{Kinetic Energy} = \text{Potential Energy} = \text{Force} \times \text{Distance}$$

$$= \text{Gravitational Force of Attraction} \times \text{Distance from Particle Center to Universe Center}$$

With: $v_{esc} \equiv$ escape velocity

$m_p \equiv$ mass of the particle

$m_U \equiv$ mass of the Universe [after the initial, mutual annihilations]

$d_0 \equiv$ distance [from the particle center of mass to the universe center of mass]

$G \equiv$ gravitation constant [un-decayed original value at the time of the "Big Bang"]

Then:

$$\tfrac{1}{2} \cdot m_p \cdot v_{esc}^2 = G \cdot \left[\frac{m_p \cdot m_U}{d_0^2} \right] \times d_0$$

$$v_{esc} = \left[\frac{2G \cdot m_U}{d_0} \right]^{\tfrac{1}{2}}$$

For that formulation the needed data are: the gravitation constant, G, the mass of the universe, m_U, and the separation distance, d_0. Estimating the Mass of the Universe, m_U, proceeds by estimating the average mass density, ρ, and the volume. The universe mass is then the product of the two. The mass density of the universe, ρ, develops as follows.

Astronomical analyses treat a "critical density" of the universe, ρ_c, which is the particular value of the average density that is on the boundary separating the case of an open (expanding forever) versus closed (eventually gravitationally recontracting) universe. The critical density relates to the escape velocity presented in equation 14-5, above. The development begins with equating kinetic and potential energy in the form of the next to last line of that equation as in equation 14-6, below.

(14-6)

$$\tfrac{1}{2} \cdot m_p \cdot v^2 = G \cdot \left[\frac{m_p \cdot m_U}{d} \right]$$

The "Hubble Law" states that the velocity of an astral object is proportional to its distance. That law, where H_0 is the "Hubble Constant", is

(14-7) $\quad v = H_0 \cdot d$

The total mass inside a sphere of radius d is

SECTION 14 - THE COSMOS' EXPANSION FROM THE ORIGIN TO THE PRESENT

(14-8) $\quad M = [\text{Volume}] \cdot [\text{density}] = [4/3 \cdot \pi \cdot d^3] \cdot [\rho]$

Substituting in equation 14-6 for v with equation 14-7 and for m_U with equation 14-8 the result is as follows.

(14-9)
$$\tfrac{1}{2} \cdot m_p \cdot [H_0 \cdot d]^2 = G \cdot \left[\frac{m_p \cdot [[4/3 \cdot \pi \cdot d^3] \cdot \rho]}{d} \right]$$

$$\rho = \frac{3 \cdot H_0^2}{8 \cdot \pi \cdot G} \qquad \text{[Simplifying and solving for } \rho\text{]}$$

That formulation is intended to give the average density of a portion of the universe of volume $4/3 \cdot \pi \cdot d^3$ such that the mass is on the boundary between escape from that volume and ultimate recapture. It would also, then, be the critical average density, ρ_c, for the overall universe, except for the following problem.

The very concept of the "Hubble Constant" is only valid in terms of the Hubble - Einstein theory that it is space itself that is expanding. It is that which would, if valid, justify the concept of one number, a "universal constant", representing the ratio of distance to velocity. The analogy given for the Hubble - Einstein concept of H_0 is that of the blowing up of a balloon or the rising of a loaf of bread in both of which examples the separation velocity of two locations within is proportional to their separation distance.

However, the Hubble - Einstein cosmology and its "Hubble Constant" concept are not valid, as already presented. The *form* of equation 14-9 is valid and correct, but the constant, H_0, must be replaced with a valid number, the correct ratio of distance to velocity for the object the escape of which is being considered, and that number is not a constant but, rather, depends on the particular circumstances.

Even in Hubble - Einstein terms, the "Hubble Constant", H_0, would better be referred to as the "Hubble Parameter". Not even the first digit of its numerical value is securely determined and its value has been taken to be over a range of from less than $H_0 = 50$ to nearly $H_0 = 100$ for various calculations and estimates by various researchers.

Further, in the "Hubble Law", $v = H_0 \cdot d$, the distance d is the distance of the astral object from the *observer*. The correct distance for the form of equation 14-9, that is the distance as in Universal Decay terms not Hubble - Einstein terms, is the distance *outward from the origin* of the "Big Bang". In other words, the "law" is that the object's outward velocity from the origin of the "Big Bang" must be, and must have been, relatively faster if its distance outward, the time-integral of that velocity, is greater, which is obvious. The "Hubble Law" is correct to that extent, but only to that.

Of course the Hubble - Einstein cosmology involves even greater error in attributing redshifts solely to the Doppler Effect of the astral object's velocity rather than the dominant cause, the Universal Decay. That means that determinations of the distance of astral objects by taking their outward velocity from the redshift as a purely Doppler effect, an incorrect velocity, and multiplying it by H_0, an invalid number and concept, can produce only distances in error.

Because for lesser distances from now back into the past [perhaps to 4 or 5 Gyrs ago] Hubble - Einstein redshift calculations of distance deviate relatively less from the correct Universal Decay calculations [Figure 14-8] a look at the results given by equation 14-9 may nevertheless be somewhat helpful in estimating universe average mass density. Depending on the value of H_0 used in equation 14-9, various values for the mass density ρ result, for example:

Value of ρ with the now favored H_0 = 72 km/sec/mpc:

$$\rho = 9.8 \cdot 10^{-27} \text{ kg/}_{m^3}$$

Value of ρ with the past favored H_0 = 49 km/sec/mpc:

$$\rho = 4.5 \cdot 10^{-27} \text{ kg/}_{m^3}$$

On the other hand estimates of ρ, rather than theoretical calculations as just above, have been made by estimating the mass of a typical galaxy, that done by estimating the number of stars in a galaxy and multiplying by the estimated average star mass and considering the galaxy's rotational dynamics; then counting the number of galaxies in a volume of space, the process performed for increasingly larger volumes. That procedure has produced a universe mass density estimate of:

Value of ρ from estimating star mass densities:

$$\rho \approx 10^{-27} \text{ kg/}_{m^3}$$

Having, then, estimates ranging from about 1 to 10 times 10^{-27}, a reasonable value to use for the mass density of the universe would be the average, about:

(14-10) $\quad \rho_U \approx 5 \cdot 10^{-27} \text{ kg/}_{m^3}$

Next the volume of the universe is needed so as to obtain the universe's mass as the product of the mass density and the volume. The <u>volume of the universe</u> develops as follows.

The particles of matter of the universe cannot have commenced their travel outward from the origin of the "Big Bang" at one same speed; rather their initial speeds must have been over a range of speeds, which would have produced a wide distribution in space as their travel developed. While the preceding analysis has developed an *average* mass density for the universe, ρ, the actual density must vary substantially even on the scale of large volumes. Therefore, to address the issue of to what volume the average mass density is to be applied requires addressing the issue of the distribution of the initial velocities of the matter emerging from the "Big Bang" because that velocity distribution is the cause of the spatial distribution of astral objects.

The analysis further on below related to Figure 14-2c, *First Phase of The Expansion of The Universe -- Velocities for Age = 30 Gyrs Case*, shows that the limits on the range of initial energies of those emerging particles set that range to energies of about

SECTION 14 - THE COSMOS' EXPANSION FROM THE ORIGIN TO THE PRESENT

`0 to 3,000 x [the escape energy]`. Those limits are the obvious lower limit of zero and an upper limit of energy so great that the matter fails to slow to non-relativistic speed ever. However, that is a very large range. Even to only `1,000` is quite large. The range used here for sample cases will be from `1 to 1,000 x [the escape energy]`. We cannot know the exact distribution of those energies; however, there are known energy distributions of other natural phenomena that can be a guide.

Those considered are Planck Black Body Radiation and the Maxwell - Boltzman treatment of the kinetic theory of gases. Replacing the case-specific constants [π, h, c, k, *2, and parameter T*] with summary case-neutral constants the form of those distributions is as in equation 14-11, and they appear as in Figures 14-1a and 14-1b, below.

(14-11) `Where:`
`F = multiple Factor of escape energy`
`n(F) = the number of particles of energy multiple F`
`p(F) = the probability of interval [F+ΔF], ΔF→0`

Planck:
$$n(F) = \frac{K1 \cdot F^5}{\varepsilon^{K2 \cdot F} - 1}$$

Maxwell-Boltzman:
$$p(F) = \frac{K3 \cdot F^{1/2}}{\varepsilon^{K4 \cdot F}}$$

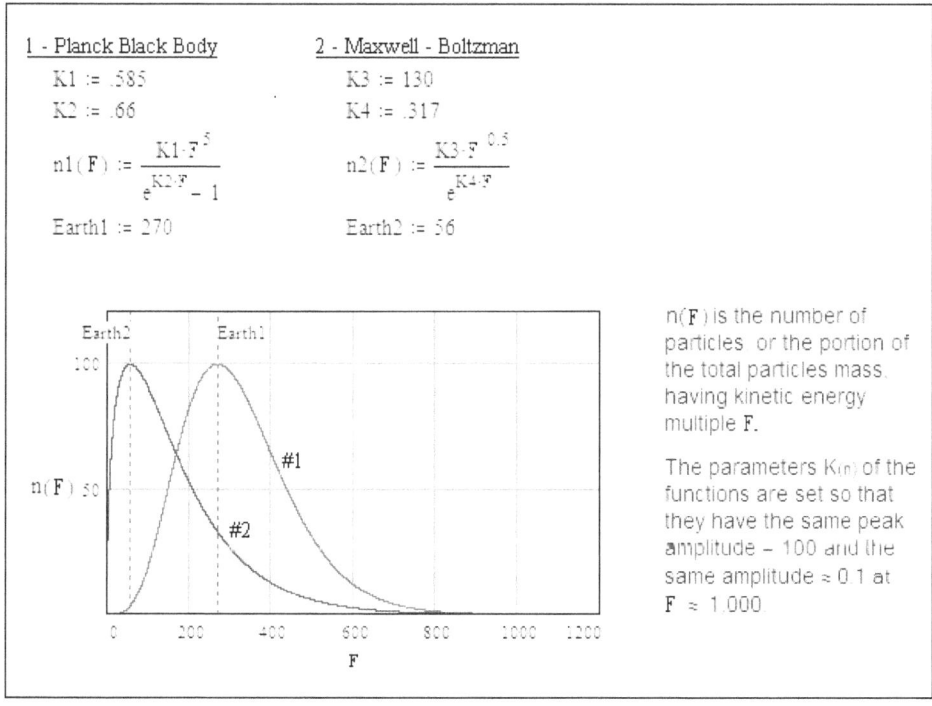

Figure 14-1a "Big Bang": Some Theoretical Rate Distributions of Initial Particle Energies

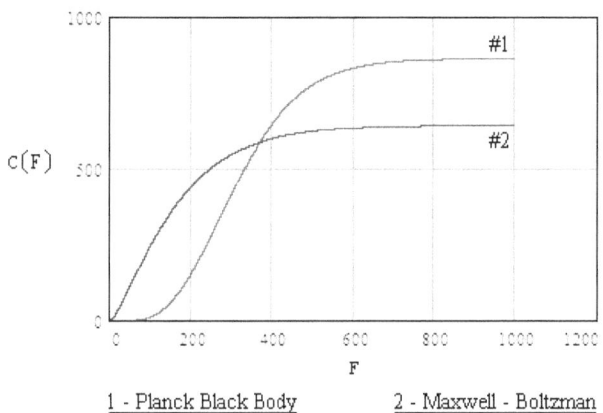

1 - Planck Black Body   2 - Maxwell - Boltzman

Figure 14-1b Cumulative Distributions for Figure 14-1a

Where we observers on planet Earth fall in the distributions in the above Figure 14-1a should be considered. It can only be presumed that Earth is not unusual with regard to its component particles' initial velocities, which would call for placing it at a distribution peak. But, there are two choices shown, likely neither exact, and Earth is not necessarily so usual as to fall exactly at the peak of any distribution. Form #2 requires less total energy and also is chosen because otherwise the resulting velocities at `Age = 30 Gyrs` would appear to be in error relative to the known velocity of Earth (see before equation 14-24, further below. Calculating for the two cases in Figure 14-1a shows that the variation in choices produces little variation in the overall results and in the age, `30 Gyrs`, of the universe and the theoretical limit, about `27 Gyrs`, on how far back in the past can be observed. The parameter for the Earth case, `F = 55`, [see Tables 14-2a and 14-2b, further below] is chosen to set it at the distribution Form #2 peak.

From the above the summary conclusions in Table 14-1c, below, can be drawn.

Form #1 -- Planck Black Body Radiation			
Percent of Maximum Amplitude, $n(F)$:	`100 %`	`95 %`	`90 %`
Range in Distribution is $F = 0$ to:	`1,000`	`565`	`500`
That Range as Percent of the Maximum	`100 %`	`57 %`	`50 %`
Resulting Indicated Universe Radius	`~14 G-Lt-Yrs`	`~14 G-Lt-Yrs`	`~14 G-Lt-Yrs`
Form #2 -- Maxwell - Boltzman Gas Kinetic Theory			
Percent of Maximum Amplitude, $n(F)$:	`100 %`	`95 %`	`90 %`
Range in Distribution is $F = 0$ to:	`1,000`	`455`	`360`
That Range as Percent of the Maximum	`100 %`	`46 %`	`36 %`
Resulting Indicated Universe Radius	`~14 G-Lt-Yrs`	`~14 G-Lt-Yrs`	`~14 G-Lt-Yrs`

Table 14-1c "Big Bang" Initial Energy Distributions Summary Data Conclusions

SECTION 14 - THE COSMOS' EXPANSION FROM THE ORIGIN TO THE PRESENT

In the above table, the "*Resulting Indicated Universe Radius*" is obtained as follows. Figure 14-4d, *Second Phase of The Expansion of The Universe -- Distances for Age = 30 Gyrs Case*, farther on below indicates a present [at `Age of the universe ≈ 30 Gyrs`] radius of the matter-containing volume of the universe as about `8 G-Lt-Yrs`. However, the radius applicable to the above obtained universe mass density should be based on an earlier time because the investigations into estimating that density had to treat astral objects which we observe as they were some time in the past: their distance from us divided by the speed of their light. Taking those earlier times as having been in the range of `0 to 7-8 Gyrs` into the past, which corresponds to volumes in the ratio to each other of the cube of those distances as `[0, 1, 8, 27, 64, 125, 216, 343, 512]`, and cumulatively in ratio as `[0, 1, 9, 36, 100, 225, 441, 784, 1296]` then it is reasonable to take the applicable universe radius as that which existed at the time into the past corresponding to about half the maximum cumulative volume, `t ≈ 6.5 Gyrs` ago. Figure 4d indicates the radii given in the above Table 14-1c for the related table columns at that time ago, `≈ 14 G-Lt-Yrs`.

Then, the estimated radius of the universe for the present calculation is:

(14-12) $R_U = 14$ `G-Lt-Yrs`

$\qquad\qquad = 11 \cdot 10^{24}$ meters.

Therefore the mass of the universe, as the product of its volume based on that radius and its equation 14-10 density, is:

(14-13) $M_U = 3 \cdot 10^{49}$ kg.

[Calculating with alternative values for the mass of the universe ranging from 10^{46} to 10^{53} `kg` produces no significant change in the general results developed below as can be verified using the forms of the calculations presented farther on below. That is, while velocities and distances vary somewhat, the necessary age of the universe remains at about the `30 Gyrs` and the maximum distance back into the past that it is theoretically possible to observe remains at about the `27 Gyrs` developed farther on below.]

[A possible concern over circular cause and effect reasoning here is not valid. The results presented are based on numerous iterations of calculations over a range of complexly interacting variables.]

With the mass of the universe now resolved the other quantities needed to calculate the escape velocity of the universe can be addressed. The <u>Separation Distance</u>, d_0, is the radius of the universe at the moment that expansion began being at a rate consistent with the long term development of the universe as compared to its initial more rapid [essentially instantaneous] development commonly referred to as "inflation". That value is $d_0 = 4.0 \cdot 10^7$ `meters`.

$G(0)$, the <u>Gravitation Constant</u> at its original un-decayed value at the time of the "Big Bang" is as follows.

(14-14)
$$G(t) = G(0) \cdot \varepsilon^{-3 \cdot t/\tau} \quad \text{meter}^3/\text{kg-sec}^2 \quad \text{[grav'n constant}$$
$$\text{universal decay}$$
$$\text{and the [}meter^3\text{] requires } \tau \to \tau/3]$$

$$G(Age) = 6.672,59 \cdot 10^{-11} \quad m^3/kg \cdot s^2 \quad [\text{"Our" G, now}]$$

$$G(Age) = G(0) \cdot \varepsilon^{-3 \cdot Age/\tau} \quad [\text{Set } t = Age \text{ in } G(t)]$$

$$G(0) = G(Age) \cdot \varepsilon^{+3 \cdot Age/\tau} \quad [\text{Solve for } G(0)]$$

$$= 6.672,59 \cdot 10^{-11} \cdot \varepsilon^{3 \cdot Age/\tau} \quad m^3/kg \cdot s^2$$

Then, the gravitation constant at any arbitrary time, t, after the "Big Bang" for any arbitrary age of the universe, Age, is as follows.

(14-15)
$$G(t, Age) = G(0) \cdot \varepsilon^{-3 \cdot t/\tau}$$
$$= [6.672,59 \cdot 10^{-11} \cdot \varepsilon^{3 \cdot Age/\tau}] \cdot \varepsilon^{-3 \cdot t/\tau} \quad m^3/kg \cdot s^2$$

Two values for the Age of the universe are addressed in this analysis to present the thesis and its validation. The currently accepted values in Hubble - Einstein cosmology range $Age = 13.5$ to 14.7 $Gyrs$. Representing those 14.0 $Gyrs$ will be used. As developed below, the present analysis indicates that $Age = 30.0$ $Gyrs$. Then, using $\tau = 11.3373$ $Gyrs$ from Equation 14-1 the following values for $G(0)$ result.

(14-16)   For Age = 14 Gyrs   For Age = 30 Gyrs
$$G(0) = 2.711 \cdot 10^{-9} \quad\quad G(0) = 1.870 \cdot 10^{-7}$$

The escape velocity per equation 14-5 for those cases of age of the universe are:

(14-17)   For Age 14 Gyrs   For Age 30 Gyrs
$$v_{esc} = 6.4 \cdot 10^{16} \, m/s \quad\quad v_{esc} = 5.3 \cdot 10^{17} \, m/s$$

Those values are so large relative to the speed of light at the time of the "Big Bang",

(14-18)   For Age 14 Gyrs   For Age 30 Gyrs
$$c(0) = 1.031 \cdot 10^9 \, m/s \quad\quad c(0) = 4.227 \cdot 10^9 \, m/s$$

that it is certain that the initial particle velocities, at the time of the "Big Bang", were very nearly the then speed of light. That is, the initial particle velocities could not be, nor exceed, light speed as the non-relativistically calculated escape velocities of equation 14-17 call for. The accommodation to relativity means that the actual speeds were very near light speed and the masses were significantly relativistically increased.

SECTION 14 - THE COSMOS' EXPANSION FROM THE ORIGIN TO THE PRESENT

As noted earlier, to determine the travel of cosmic bodies outward from the "Big Bang" one needs to know, first, the initial velocities and then the subsequent manner of the reduction in the cosmic bodies' velocities by gravitation and other effects. The initial velocities have been found to be essentially the value of the speed of light at the time of the "Big Bang". At that point two effects proceeded to slow the outward velocities: the decay of the speed of light [the upper limit on particle velocity] and the gravitational slowing [the centrally directed gravitational acceleration caused by the total mass].

(2) The Progressive Reduction in the Cosmic Bodies' Initial Velocities

The overall process must be divided into two phases:

- First, the relativistic phase during which the speed is continuously almost that of light and the effect of gravitation is dominantly not a reducing of the speed but a reducing of the amount that the mass has been relativistically increased, and

- Second, the non-relativistic phase during which the mass, now reduced to essentially rest mass, remains essentially the same and the dominant effect of gravitation is to reduce the speed.

Of course the change from the first to the second phase is not sharp, but rather a gradual smooth transition. For the purposes of these calculations, however, the choosing of a specific transition point [hereafter termed the `ChangePoint`] is needed. That point is determined as follows.

(2a) The First Calculation Phase -- Speed ≈ Light Speed

The first phase calculation is in terms of energy, the gradual transfer of initial kinetic energy into gravitational potential energy. Energy calculations in themselves are not relativistic. The kinetic energy speed will be treated non-relativistically, that is, the mass is taken as at its rest value and the kinetic energy then is taken as all residing in the [theoretical] velocity squared, that theoretical velocity not constrained by a speed of light limitation. Then, the calculated effect of gravitation, of the transfer of kinetic energy into gravitational potential energy, appears as a gradual reduction of that theoretical velocity. When that theoretical velocity has been reduced by gravitation down to the actual [at that time as decayed] light speed then the `ChangePoint` from the relativistic to the non-relativistic treatment has been reached.

During that first phase the distance component of the gravitational potential energy calculation is readily available as the time integral of the known speed, the speed of light. The velocity as a function of time then develops as follows.

The first phase distance, $d(t, Age)$, traveled outward from the "Big Bang" source location, as a function of time is the time integral of the velocity as equation 14-19, below, which is based on equation 14-7, above [and includes the initial separation distance, $d_0 = 4.0 \cdot 10^7$ meters, of the earlier above calculation of the mass of the universe, which distance is negligible, however].

$$(14\text{-}19) \quad d(t, Age) = d_0 + \int_0^t c(t) \cdot dt$$

$$= d_0 + \int_0^t \left[2.997,924,58 \cdot 10^8 \cdot \varepsilon^{Age/\tau} \right] \cdot \varepsilon^{-t/\tau} \cdot dt$$

The velocity as a function of time, $v(t, Age)$, is as given in equation 14-20 obtained starting from equation 14-5:

$$(14\text{-}20) \quad v_{esc} = \left[\frac{2G \cdot M_U}{d_0} \right]^{1/2} \quad \text{so that:}$$

$$v(t, Age) = \left[\frac{2G \cdot M_U}{d(t, Age)} \right]^{1/2}$$

As pointed out during the evaluation of the mass of the universe earlier above, the matter of the universe moved outward from the "Big Bang" at a wide range of speeds. Those various speeds resulted, of course, from the particles of matter having various initial velocities / energies which, as presented just before and in conjunction with Figures 14-1a and 14-1b, are to be sampled over the range $F = 1$ to $1,000 \times$ [the escape energy]. That range is incorporated into the formulation by the multiple factor, F, included in the final expression for the first phase $v1(t, Age)$ per below.

$$(14\text{-}21) \quad v1(t, Age) = \left[\frac{\mathbf{F} \cdot 2G \cdot M_U}{d(t, Age)} \right]^{1/2} \quad \text{[1st Phase Matter Velocities]}$$

The decaying speed of light is per equation 14-3, repeated below,

$$(14\text{-}22) \quad c(t, Age) = \left[2.997,924,58 \cdot 10^8 \cdot \varepsilon^{Age/\tau} \right] \cdot \varepsilon^{-t/\tau}$$

and the value of time, t, producing $\boxed{v1(t, Age) \equiv c(t, Age)}$ is the sought *ChangePoint* for the particular initial velocity / energy multiple factor, F, and Age, the end of calculation for the first, the relativistic, phase.

Tables 14-2a and 14-2b, below summarize the results for the first phase for both $Age = 30$ and $Age = 14$ $Gyrs$, and the results are also presented graphically for $Age = 30$ $Gyrs$ in Figure 14-2c, following the tables.

SECTION 14 - THE COSMOS' EXPANSION FROM THE ORIGIN TO THE PRESENT

```
For:  Universe Age = 30 Gyrs, which means that:
      Initial Light Speed = 4,226,895.62 ·10^9 m/s
      Initial Gravitation Constant, G = 1.870,24 ·10^-7 m^3/kg-s^2
```

F-Factor	At Time [Gyrs]	At Velocity [m/s]	"ChangePoint" Distance From Origin* ChangePoint	Now, Age	Relative! Abundance
1	0.004713	4.225·10^9	0.066	0.005	22.
3	0.01414	4.222·10^9	0.199	0.014	37.
10	0.04721	4.209·10^9	0.661	0.047	63.
32	0.1518	4.171·10^9	2.097	0.152	93.
55 Earth	0.2621	4.131·10^9	3.568	0.262	100.
100	0.4812	4.052·10^9	6.364	0.481	90.
316	1.5960	3.672·10^9	18.222	1.596	23.
1000	6.0840	2.472·10^9	38.788	6.084	0.1
≈3000	→ ∞	c(t)	→ ∞	→ ∞	

* = Decayed to Change Point, Age; G-Lt-Yrs.   ! = Estimate per Fig 14-1a

Table 14-2a
The Universe's First Phase of Expansion, Age=30 Gyrs Case
Relativistic Phase at (Essentially) Light Speed, From t = 0 to t = ChangePoint

```
For:  Universe Age = 14 Gyrs, which means that:
      Initial Light Speed = 1,030,357.62 ·10^9 m/s
      Initial Gravitation Constant, G = 2.711,29 ·10^-9 m^3/kg-s^2
```

F-Factor	At Time [Gyrs]	At Velocity [m/s]	"ChangePoint" Distance From Origin* ChangePoint	Now, Age	Relative! Abundance
1	0.004715	1.030·10^9	0.016	0.005	22.
3	0.01416	1.029·10^9	0.049	0.014	37.
10	0.04721	1.026·10^9	0.161	0.047	63.
32	0.1518	1.017·10^9	0.511	0.151	93.
55 Earth	0.2621	1.007·10^9	0.070	0.259	100.
100	0.48125	9.878·10^8	1.551	0.471	90.
316	1.5961	8.953·10^8	4.443	1.488	23.
1000	6.089	6.024·10^8	9.460	4.708	0.1
≈3000	→ ∞	c(t)	→ ∞	→ ∞	

* = Decayed to Change Point, Age; G-Lt-Yrs.   ! = Estimate per Fig 14-1a

Table 14-2b
The Universe's First Phase of Expansion, Age=14 Gyrs Case
Relativistic Phase at (Essentially) Light Speed, From t = 0 to t = ChangePoint

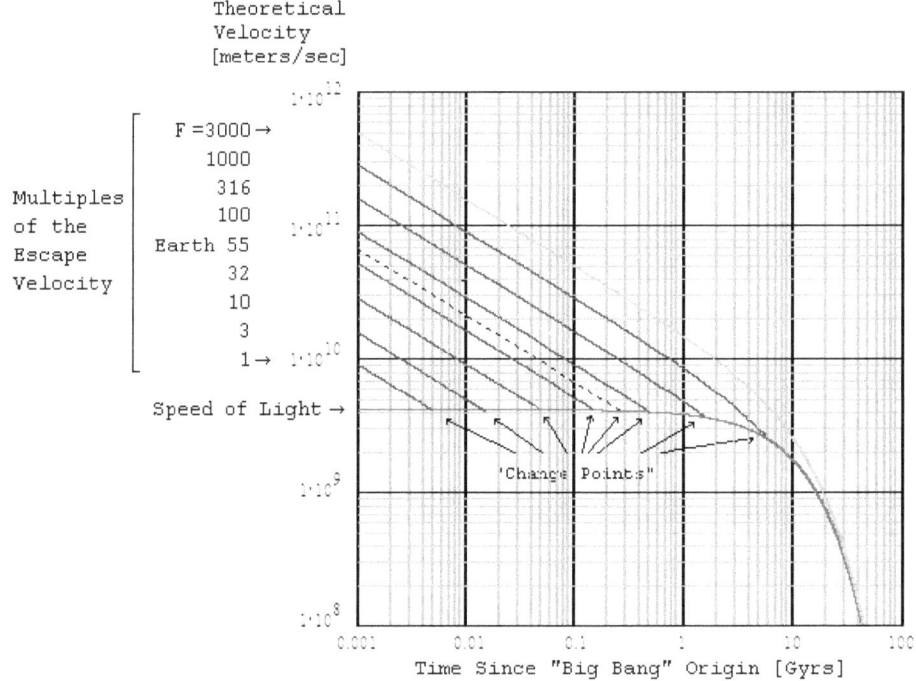

Figure 14-2c First Phase of The Expansion of The Universe
Velocities for Age = 30 Gyrs Case

Note that for values of the `F-Factor` at about `F = 3,000` and above the "*ChangePoint*" is never reached because of the decay in the speed of light. For those values the outward moving matter never slows below, essentially, the then on-going decaying light speed.

Note, also, that the large Doppler red shift resulting from v nearly equaling c combined with large decay redshift due to lack of much decay because the time lies only shortly after $t = 0$ results in a redshift relative to our "normal" local wavelengths by a factor of `24 - 28`. The least wavelength of visible light is about `0.38 microns`. That wavelength shifted by that factor as `0.38 x [24 - 28] = [9 - 11] microns`, lies well into the infra-red portion of the spectrum.

Consequently, light which would otherwise lie in the "visible light" portion of the spectrum but was emitted from sources before they reached their "ChangePoints" lies shifted sufficiently into the infra-red that its detection is relatively unlikely. That is the case especially because sources with relatively later "ChangePoints" [more recent, therefore more susceptible to observation] are of relatively small relative abundance. In other words, light emitted from astral sources before they reached their `ChangePoint` is much less likely to be observed.

SECTION 14 - THE COSMOS' EXPANSION FROM THE ORIGIN TO THE PRESENT

(2b) The Second Calculation Phase -- Speed < Light Speed

Gravitational slowing is an awkward problem. The amount of gravitational slowing depends on the distance outward from the origin of the "Big Bang"; those distances depend on the velocity function during the travel from the origin outward; and that velocity function depends on the gravitational slowing -- a problem of circular cause and effect. The calculation breaks down into two different modes of behavior because of relativistic effects.

The first phase of the outward expansion, already analyzed above, takes place at essentially the actual speed of light regardless of gravitation. That is because the [theoretical non-relativistic] escape velocities are so large. The kinetic energy essentially resides in the relativistically increased mass until gravitation has reduced the [theoretical non-relativistic] greater than light speed down to passing through and below the actual light speed. During that first phase distances are known because the velocities are known independently of the distance; they are essentially the then current, as decayed, light speeds.

The second phase begins at the end of that first phase's known outward travel to its "*ChangePoint*". At that point the circular cause and effect awkwardness of the problem returns. The inverse square gravitational behavior calls for the current total outward distance squared in its denominator and that depends on the velocity history which depends on the distance history which depends on the velocity history. The solution is to use an approximating function of similar form but not involving velocity. That function can then be adjusted by calibrating it to the known velocity of the Earth now, as developed further below.

The <u>approximating function</u> develops as follows. The precise behavior of the universe's matter expanding outward from the "Big Bang" is as equation 14-23, below, the distance represented by the variable s to avoid confusion with the symbols for differentiation.

(14-23)
$$\text{Gravitational Deceleration} = \frac{d^2 s}{dt^2} = - \frac{G \cdot \text{UniverseMass}}{s^2}$$

The general form of the solution to that equation is as equation 14-24:

(14-24)
$$\text{Velocity} = \frac{ds}{dt} = \frac{1}{A \cdot \varepsilon^{B \cdot s} + C}$$

which states that the velocity is inversely proportional to the exponential of the distance, s, as $1/\varepsilon^s$. The procedure will be to use as the approximation to the actual exact velocity function the function for the speed of light, $c(t)$, per equation 14-22, multiplied by a factor based on $1/\varepsilon^s$.

However, the velocity function must be in terms of time, t, as the independent variable, not distance, s, else the problem of circular cause and effect remains. Using t instead of s, that is representing the actual exact velocity function with a function multiplying the speed of light, $c(t)$, by a factor based on $1/_\varepsilon t$ resolves that problem but is less accurate an approximation. The problem of accuracy is addressed by calibrating to the known velocity of our planet Earth at time the Age of the universe.

Doppler analysis of the cosmic microwave background radiation, the Doppler variation being due to the Earth's orbit around the Sun, shows that the absolute velocity of the Earth [absolute relative to the location of the "Big Bang" origin] is now about $3.7 \cdot 10^5$ $meters/_{sec}$. The calibration of the velocity approximating function must be such as to produce approximately that velocity now, at time $t = Age$ after the "Big Bang"; but, for what case, what value of the F-$Factor$? That issue has been already addressed above and the result is that, for $Earth$ $F = 55$ will be used.

For the beginning of the second phase to match smoothly with the end of the first phase the adjustment component, $1/_\varepsilon t$, which increases in its effect as t increases, must be formulated so as to produce zero change when the time is $t = ChangePoint$. The resulting expression for the second phase velocity is equation 14-25, below.

(14-25)
$$v2(t, Age) = c(t, Age) \cdot \frac{1}{_\varepsilon A \cdot (t - ChangePoint)} \quad meters/_{sec}$$

where:
  A = a constant of value yet to be determined,
     chosen to calibrate the function, and
  t ≥ ChangePoint.

The calibrating constant, A, for the case of $Earth$, F-$Factor = 55$ is set to the value that produces a velocity at Age of $3.7 \cdot 10^5$ $meters/_{sec}$, the known value as already presented. The value of A for the highest F-$Factor$ case is set to produce a velocity of the current speed of light, $3 \cdot 10^8$ $meters/_{sec}$, at Age. The value of A for each of the other cases is interpolated using a decaying exponential form to provide a general representative set of samples. That form is per equation 14-26, below, and is derived from its depiction in Figure 14-3 *Calibrated Velocities at "Age" for Sample Expansion Cases #1 - 7, and #Earth*, below.

(14-26)
$$Velocity(i) = 5.452 \cdot 10^9 \cdot [\varepsilon^{-2.2127(9-i)}]$$
$$\text{where } i = \text{case } \# = 1, 2, \ldots 8$$

The development of the formulations for the distances is presented further below at equations 14-29A through 14-31B.

SECTION 14 - THE COSMOS' EXPANSION FROM THE ORIGIN TO THE PRESENT

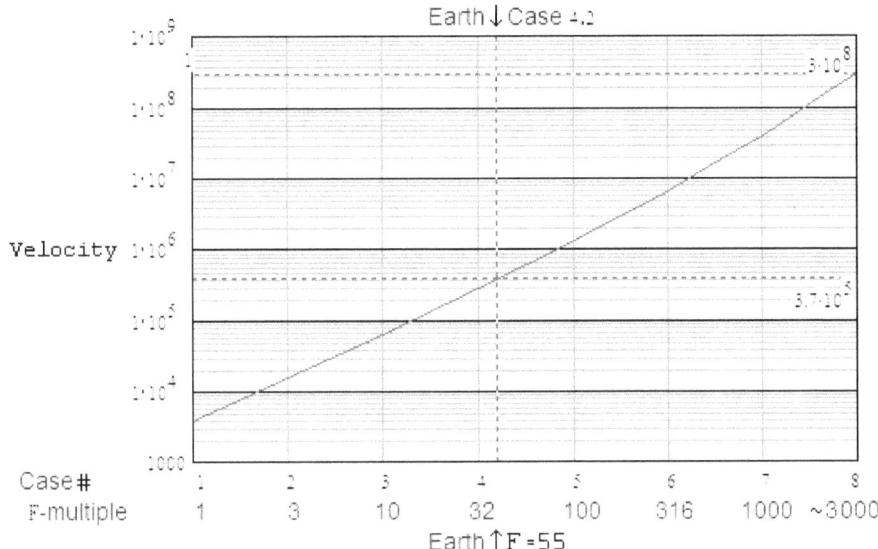

Figure 14-3
Calibrated Velocities at "Age" for Sample Expansion Cases #1 - 7, and #Earth

Tables 14-4a and 14-4b, below, summarize the results for the second phase for *Age = 30 Gyrs* and for *Age = 14 Gyrs*, and the results are also presented graphically for *Age = 30 Gyrs* in Figure 14-4c, *Second Phase of The Expansion of The Universe -- Velocities for Age = 30 Gyrs Case*, and Figure 14-4d, *Second Phase of The Expansion of The Universe -- Distances for Age = 30 Gyrs Case*, on the page following the tables.

	For: Universe Age = 30 Gyrs, which means that: Initial Light Speed = $4.226,895,62 \cdot 10^9$ m/s Initial Gravitation Constant, G = $1.870,24 \cdot 10^{-7}$ m³/kg-s²			
F-Factor	ChangePoint Time [Gyrs]	2nd Phase Constant-A	At Age = 30 Gyrs, [Data*] Velocity [m/s]	Distance [G-Lt-Yrs]
1	0.004713	16.92385	$0.00003814 \cdot 10^8$	2.157
3	0.01414	16.98010	$0.0001505 \cdot 10^8$	2.400
10	0.04721	17.04691	$0.0006230 \cdot 10^8$	2.727
32	0.1518	17.12856	$0.002731 \cdot 10^8$	3.205
55	0.2621	17.14622	$0.003700 \cdot 10^8$	3.374
100	0.4812	17.23245	$0.01287 \cdot 10^8$	3.980
316	1.5960	17.37183	$0.06673 \cdot 10^8$	5.388
1000	6.0840	17.57200	$0.3974 \cdot 10^8$	8.034
≈3000	→ ∞	n/a	$2.99792458 \cdot 10^8$	10.526
				* = Decayed to Age

Table 14-4a
Summary Data Results for the Universe's Second Phase of Expansion
Non-Relativistic Phase From t = "ChangePoint" Onward, Age = 30 Gyrs

93

For:	Universe Age = 14 Gyrs, which means that: Initial Light Speed = $1.030,357,62 \cdot 10^9$ m/s Initial Gravitation Constant, $G = 2.711,29 \cdot 10^{-9}$ m^3/kg-s^2			
F-Factor	ChangePoint Time [Gyrs]	2nd Phase Constant-A	At Age = 14 Gyrs, [Data*] Velocity [m/s]	Distance [G-Lt-Yrs]
1	0.004715	16.59278	$0.00003814 \cdot 10^8$	1.122
3	0.01416	16.64887	$0.0001505 \cdot 10^8$	1.268
10	0.04721	16.71513	$0.000623 \cdot 10^8$	1.477
32	0.1518	16.79505	$0.002731 \cdot 10^8$	1.811
55	0.2621	16.81082	$0.003700 \cdot 10^8$	1.954
100	0.4812	16.89330	$0.01287 \cdot 10^8$	2.417
316	1.5961	17.01200	$0.06673 \cdot 10^8$	3.669
1000	6.0890	17.09155	$0.3974 \cdot 10^8$	6.295
≈3000	→ ∞	n/a	$2.99792458 \cdot 10^8$	8.034

* = Decayed to Age

Table 14-4b
Summary Data Results for the Universe's Second Phase of Expansion Non-Relativistic Phase From t = "ChangePoint" Onward, Age = 14 Gyrs

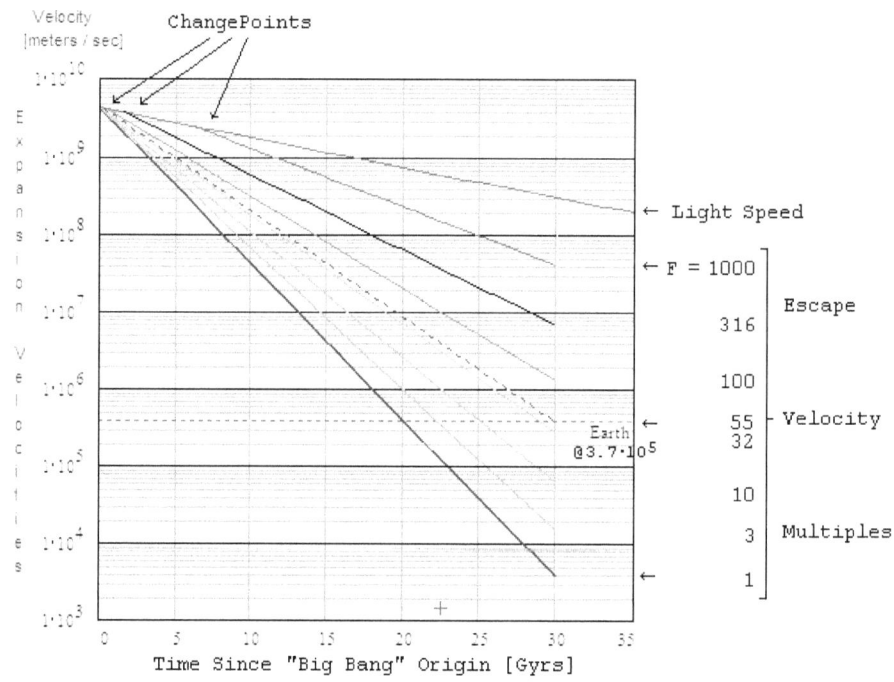

Figure 14-4c
Second Phase of The Universe Expansion -- Velocities for Age = 30 Gyrs Case

SECTION 14 - THE COSMOS' EXPANSION FROM THE ORIGIN TO THE PRESENT

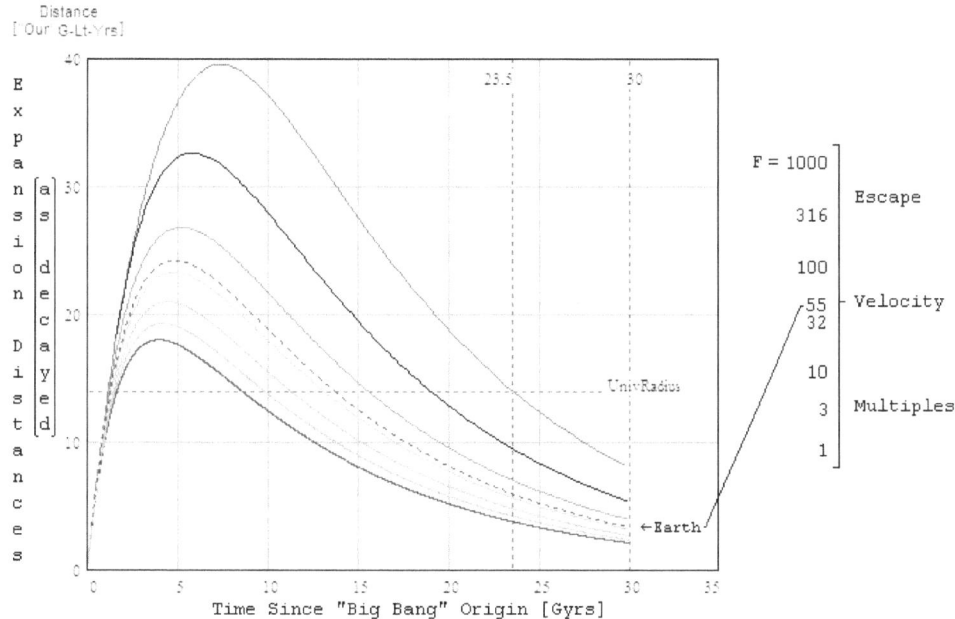

Figure 14-4d
Second Phase of The Universe Expansion -- Distances for Age = 30 Gyrs Case

[The notation "UnivRadius" in the above Figure 14-4d refers to the discussion following Table 14-1c concerning developing the appropriate volume of the universe to use in conjunction with the universe density to obtain the universe mass.]

2 - OBSERVING, FROM EARTH, LIGHT EMITTED BY ASTRAL SOURCES

Now that formulations have been developed that describe the travel of the two different traveling entities, light and matter, at various times in the past from at the beginning to the present the problem of observing light from astral sources can be addressed. The problem is to determine under what conditions light emitted long ago will have traveled to the exact present location of an observer that has also been performing its own travel while the light to be observed has been traveling.

For convenience the following quantities are defined.

```
Age  ≡ Age now of the universe = time "now".
Back ≡ how long ago it is theoretically possible to
       observe.
Then ≡ the Age of the universe then [Back ago].
     = Age - Back.
```

The distance that the light travels [continuously at whatever its speed was when it was first emitted from its astral source] from when first emitted at time t until now, at time *Age*, is then:

95

TTHE TROUBLE WITH THE HUBBLE LAW

(14-27)
$$\begin{aligned}
\text{LightTravel}(t, \text{Age}) &= [\text{Speed}(t)] \cdot [\text{Travel Time}] \\
&= [c(t, \text{Age}) \, [m/s]] \cdot [[\text{Age} - t] \, [s]] \\
&= c(t, \text{Age}) \cdot [\text{Age} - t]
\end{aligned}$$

Per the above mnemonic terminology, $\text{LightTravel}(t, \text{Age})$, the motion of the cosmic bodies involved will be termed as in equation 14-28.

(14-28)
$$\begin{aligned}
\text{WhereUs}(t, \text{Age}, F) &\equiv \text{location of we observers} \\
\text{WhereSource}(t, \text{Age}, F) &\equiv \text{location of observed light source} \\
&\approx -\text{WhereUs}(t, \text{Age}, F)
\end{aligned}$$

[to account for the formulations being of similar form but with their travels diametrically opposite and their "F" values able to be different].

Taking the location of the "Big Bang" singularity as at distance zero and noting that the initial distance, $d_0 = 4.0 \cdot 10^7 \text{ meters}$, is negligible [less than one "our light years" on a scale of "giga our light years"], then the formulation for $\text{WhereUs}(t, \text{Age})$ develops as follows.

(14-29)   Travel from the "Big Bang" outward

(14-29A) Travel during time t=0 to time t=ChangePoint:
$$\begin{aligned}
\text{WhereUsA}(t, \text{Age}, F) &= [\text{Travel of Matter in 1st Phase}] \\
&= \int_0^t [\text{Speed of Matter in 1st Phase}(t, \text{Age}, F)] \cdot dt \\
&= \int_0^t [\text{Decaying Light Speed}(t, \text{Age})] \cdot dt \\
&= \int_0^t c(t, \text{Age}) \cdot dt
\end{aligned}$$

(14-29B) Travel from time t=ChangePoint onward:
$$\begin{aligned}
\text{WhereUsB}(t, \text{Age}, F) &= \\
&= [\text{1st Phase Travel to ChangePoint}] \\
&\quad + [\text{2nd Phase Travel}(t, \text{Age}, F)] \\
&= [\text{WhereUsA}(\text{ChangePoint}, \text{Age})] \\
&\quad + [\text{2nd Phase Travel}(t, \text{Age}, A)] \qquad [F \to A] \\
&= [\text{1st Term}] + \int_{\text{ChangePoint}}^t [\text{Decaying } c(t)] \cdot [\text{GravSlowing}] \cdot dt \\
&= [\text{1st Term}] + \int_{\text{ChangePoint}}^t [c(t)] \cdot \frac{1}{\varepsilon A \cdot [t - \text{ChangePoint}]} \cdot dt
\end{aligned}$$

SECTION 14 - THE COSMOS' EXPANSION FROM THE ORIGIN TO THE PRESENT

where "A" is the "2nd Phase Constant-A"
of Tables 14-4a and 14-4b.

However, there is one more factor in the development of *WhereUs(t,Age)*, the general effect of the Universal Decay. The Universal Decay produces an acceleration on all bodies which acceleration is: centrally directed, independent of separation distances, and the same and constant everywhere except that the amount of that acceleration also exponentially decays. Its value now is $(8.74 \pm 0.94) \times 10^{-8}$ $cm/_s2$. That acceleration produces a gradual contraction of the overall universe; that is, the exponential decay of the length $[L]$ aspect of all quantities is also a decay of the distance spacings in the universe.

The acceleration is evidenced by galactic rotation curves and by the travel of the Pioneer 10 and 11 space craft. That the Universal Decay is not directly observable because our measuring equipment [our "ruler"] is also decaying, as noted earlier above, prevents our direct observation of the contraction in the case of our solar system and that of galactic rotation curves. However, in the case of the Pioneer 10 and 11 satellites and the case of galactic rotation curves, the decay has been detected because it forces orbital / path behavior that would not be present if there were no decay, and that behavior can be and has been observed -- e.g. the Pioneer space craft are not as far outward from the Sun as they should be were there no Universal Decay contraction.

Consequently, as the various bodies in the universe travel outward from the location of the "Big Bang", the distances that they have already traveled continuously decay. And, consequently, the distances traveled by light emitted from the various sources in the universe continuously decay after having been first traveled "undecayed". Therefore, the final form for *LightTravel* is then equation 14-29, below [continued from equation 14-26]. And, the final form for *WhereUs(t,Age)* is equations 14-30A and 14-30B, below [continued from equations 14-29A and 14-29B, above].

(14-30) <u>Distance traveled outward from its source until now, time "Age", by light emitted at time "t"</u>:

LightTravel(t,Age) = [Speed]·[Travel Time]·*[As Decayed]*

$$= c(t, Age) \cdot [Age - t] \cdot \varepsilon^{-[(Age-t)/\tau]}$$

(14-31) <u>Distance traveled outward from the "Big Bang" until time "t" by matter originating at the "Big Bang" [t=0, distance=0]</u>:

WhereUs(t,Age,F):

(14-31A):

Travel during time t=0 to time t=ChangePoint:

WhereUsA(t,Age,F) = [1st Phase Matter Travel]·*[As Decayed]*

$$= \left[\int_0^t c(t, Age) \cdot dt \right] \cdot \varepsilon^{-[t/\tau]}$$

97

(14-31B):
    Travel from time t=ChangePoint onward:
    WhereUsB(t,Age,F) =
        = [[1st Phase Travel to ChangePoint]·[not decayed]
                + [2nd Phase Travel]]·*[All as Decayed]*

$$= \left[[1st] + \int_{ChangePoint}^{t} [c(t)] \cdot \frac{1}{\varepsilon^{A \cdot [t - ChangePoint]}} \cdot dt \right] \cdot \varepsilon^{-[t/\tau]}$$

a. *The Maximum Distance into the Past That is Observable*

The extreme case of observing ancient light is the observing of light that originated diametrically opposite from us, the observers, relative to the origin of the "Big Bang". That light <u>must</u> travel the distance from its source back to the location of the origin of the "Big Bang" and then further outward to the location of us, the observers. The light originates from its source at time $t = Then$ and is observed by us at time $t = Age$. That distance, for any age of the universe, Age, is as follows.

(14-32)
    LightMustTravel(t,Age)
        = -WhereSource(Then,Age,F) + WhereUs(Age,Age,F)

As compared to the above requirement, the actual distance that that light <u>does</u> travel is given by equation 14-29 with $t = Then$, as follows.

(14-33)
    LightDoesTravel(t,Age)
        = c(Then,Age)·[Age - Then]·$\varepsilon^{[(Age - Then)/\tau]}$
        = c(Then,Age)·[Back]·$\varepsilon^{-[(Back)/\tau]}$

For the light to be theoretically observable by us the above two must be the same.

(14-34)   LightMustTravel(t,Age,F) = LightDoesTravel(t,Age)

The only variables in equation 14-34 [for a particular Age and energy multiple, F,] are $Back$ and $Then$, either of which determines the other per $Then = Age - Back$. The solution to equation 14-34 is obtained using a computer assisted design program ["Mathcad" in this case]. The applicable form of $WhereUs(t,Age,F)$ must be used, $WhereUsA(t,Age,F)$ or $WhereUsB(t,Age,F)$ depending on the value of $Then$ relative to the $ChangePoint$.

The results are presented in Tables 14-5a and 14-5b, below.

SECTION 14 - THE COSMOS' EXPANSION FROM THE ORIGIN TO THE PRESENT

```
For:  Universe Age = 30 Gyrs, which means that:
      Initial Light Speed = 4.226,895,62·10⁹ m/s
      Initial Gravitation Constant, G = 1.870,24·10⁻⁷ m³/kg-s²
```

F-Factor	ChangePoint Time [Gyrs]	2nd Phase Constant-A	Observable Past Distance Back [Gyrs]	Observable Past Distance Then [Gyrs]	Relative! Abundance
1	0.004713	16.92385	7.5860	22.4140	22.
3	0.01414	16.98010	8.4081	21.5919	37.
10	0.04721	17.04691	9.8790	20.1210	63.
32	0.1518	17.12856	26.0791	3.9209	93.
100	0.4812	17.23245	27.01869	2.98131	90.
316	1.5960	17.37183	27.56523	2.43477	23.
1000	6.0840	17.57200	27.61821	2.38179	0.1

! = Estimate per Fig 14-1a, where Earth, F = 55, is of Abundance 100

Table 14-5a
Distance into the Past That is Observable, Age = 30 Gyrs

```
For:  Universe Age = 14 Gyrs, which means that:
      Initial Light Speed = 1.030,357,62·10⁹ m/s
      Initial Gravitation Constant, G = 2.711,29·10⁻⁹ m³/kg-s²
```

F-Factor	ChangePoint Time [Gyrs]	2nd Phase Constant-A	Observable Past Distance Back [Gyrs]	Observable Past Distance Then [Gyrs]	Relative! Abundance
1	0.004715	16.59278	3.4828	10.5172	22.
3	0.01416	16.64877	3.7136	10.2864	37.
10	0.04721	16.71513	4.0702	9.9298	63.
32	0.1518	16.79505	4.6849	9.3151	93.
100	0.48125	16.89330	5.9772	8.0228	90.
316	1.5961	17.01200	8.8230	5.1770	23.
1000	6.0890	17.09155	10.04945	3.95055	0.1

! = Estimate per Fig 14-1a, where Earth, F = 55, is of Abundance 100

Table 14-5b
Distance into the Past That is Observable, Age = 14 Gyrs

For *Age = 14 Gyrs*, as presented in Table 14-5b above, even the most energetic case of observable past distance, that for *F = 1,000*, has a theoretical limit, about *10 Gyrs*, that is less than actual observations have reported [the reported distances based on Hubble - Einstein cosmology].

That Hubble - Einstein cosmology problem is even more severe if the calculations of Table 14-5b are performed with no universal decay, as the Hubble - Einstein cosmology contends. The results for that case are presented in Table 14-5c below in which the greatest observable past distance is barely *8 Gyrs*, quite substantially less than reported observations [their reported distances based, erroneously, on the Hubble Law]. That is so even when a greatly more energetic case, $F = 3000$, is examined. For that extreme the energy is such that the gravitation has not yet slowed that matter down to the speed of light which means that its Doppler redshift would be approaching the infinite, $z \approx \infty$.

```
For:  Universe Age = 14 Gyrs, and no Universal Decay
      per Hubble - Einstein Cosmology
    F-         ChangePoint    2nd Phase      Observable Past Distance     Relative!
  Factor       Time[Gyrs]     Constant-A     Back [Gyrs]    Then [Gyrs]   Abundance
     1          0.00471       16.59278         3.5544         10.4456        22.
     3          0.01413       16.64877         3.7343         10.2657        37.
    10          0.04713       16.71513         3.9972         10.0028        63.
    32          0.1518        16.79505         4.4214          9.5786        93.
   100          0.471         16.89360         5.1720          8.8280        90.
   316          1.489         17.01575         6.5490          7.4510        23.
  1000          4.710         17.16130         8.08125         5.91875        0.1
       ! = Estimate per Fig 14-1a, where Earth, F = 55, is of Abundance 100

  3000         14.13 [>Age]     n/a            8.15            5.85
```

Table 14-5c
Table 14-5b Re-Calculated with No Universal Decay
and Added Extreme Case

Clearly, the tenets of the Hubble - Einstein cosmology fail because they cannot conform to reality as it is already known.

Returning to Universal Decay cosmology and the age of the universe being $Age = 30\ Gyrs$, that age derives from what is needed to enable observation of redshifts on the order of $z = 10$, as presented in the next section below. It is an estimate because our instrumentation presently limits our ability to observe the past more than the theoretical limit does. Consequently new developments in instrumentation and observation may produce observed redshifts greater than $z = 10$, ones on the order of $z = 12$ or more, and therefore require a corresponding increase in the estimated age of the universe.

The present value for the farthest back into the past that it is theoretically possible to observe regardless of the quality of our instrumentation is a little over *27 Gyrs* ago to the time *2 to 3 Gyrs* after the "Big Bang". The travel of the light source and of the observer's home and of the emitted light for that case of the most distant source theoretically observable, all from the time of the "Big Bang" to the present are as shown in Figure 14-6, on the following page.

SECTION 14 - THE COSMOS' EXPANSION FROM THE ORIGIN TO THE PRESENT

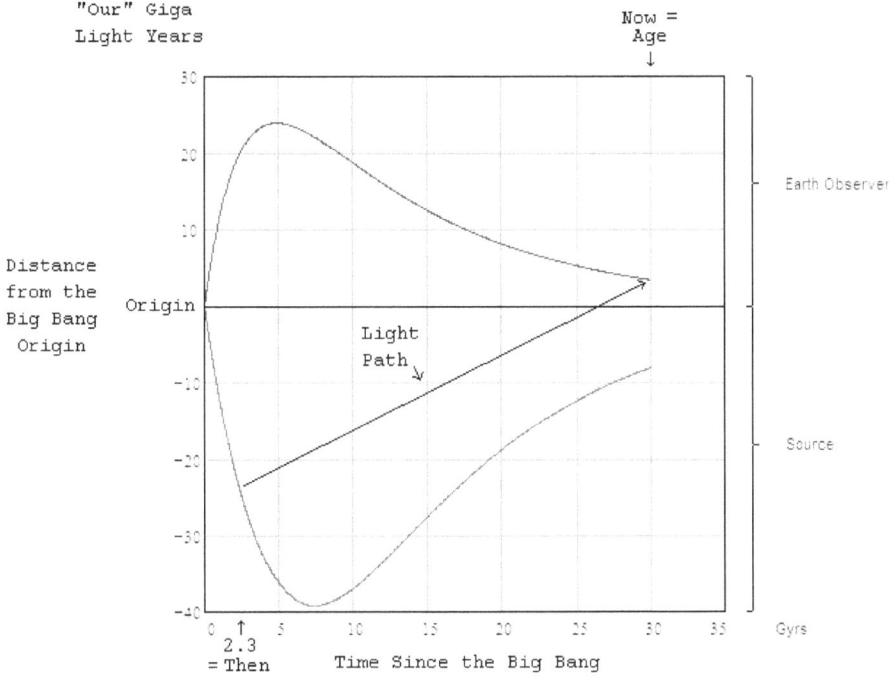

Figure 14-6
The Most Distant into The Past Source [-] Observable by an Earth Observer [+] Diametrically Opposite Relative to the "Big Bang" Origin

b. Redshifts: Universal Decay and Hubble Doppler

There are two causes of the redshifts that we observe: the universal decay and the Doppler shift due to astral objects' velocity away from us.

The universal decay redshift occurs because we observe ancient light traveling at the speed at which it was originally emitted, a speed significantly larger because less decayed than our present local speed of light. We observe the greater speed as a lengthening of all wavelengths in the light [with no change in frequencies]. The formulation for the universal decay redshift, z_τ, of light that was emitted at time $t = T$ after the "Big Bang" and is observed at a later time $t = now = age$ is as follows below.

$$(14\text{-}35) \quad z_\tau \equiv \text{redshift due to the universal decay}$$

$$= \frac{\lambda_{observed} - \lambda_{local}}{\lambda_{local}}$$

$$= \frac{c(\text{time light emitted})}{c(\text{time now})} - 1$$

$$= \frac{c(t=0) \cdot \varepsilon^{-[T/\tau]}}{c(t=0) \cdot \varepsilon^{-[age/\tau]}} - 1 = \frac{\varepsilon^{-[T/\tau]}}{\varepsilon^{-[age/\tau]}} - 1$$

The formulation for the Doppler shift due to astral objects' velocity away from us, z_D, is as follows, per standard Hubble - Einstein cosmology.

(14-36) $z_D \equiv$ relativistic redshift of the Doppler effect

$$= \frac{[1 + v/c]^{1/2}}{[1 - v/c]^{1/2}} - 1$$

The formulation for the universal decay redshift, equation 14-35, is a function of time, not velocity. Equation 14-36 can be converted to expressing the Doppler redshift, z_D, in terms of time by using the velocity-as-a-function-of-time expressions for the motion of the astral body products of the "Big Bang" developed earlier above: equations 14-21 and 14-25.

For the period from time $t = 0$ through $t =$ ChangePoint the velocity, equation 14-21, is very nearly the then current decaying speed of light. The v/c ratio is very nearly 1.0 so that the redshifts, z_D, are very large, but are also essentially meaningless for any useful purpose.

For the period from $t =$ ChangePoint onward the velocity expression is equation 14-25, repeated below.

(14-25)
$$v2(t, Age, F) = c(t, Age) \cdot \frac{1}{\varepsilon A \cdot (t - ChangePoint)}$$

where A and ChangePoint are given in Table 14-5.

The v/c ratio is equation 14-25 divided by $c(t, Age)$:

(14-37)
$$v/c = \frac{1}{\varepsilon A \cdot (t - ChangePoint)}$$

and by substituting that into equation 14-36 the expression for the Doppler redshift, z_D, is:

(14-38)

$z_D \equiv$ relativistic redshift of the Doppler effect

$$= \frac{[1 + v/c]^{1/2}}{[1 - v/c]^{1/2}} - 1$$

$$= \frac{\left[1 + \dfrac{1}{\varepsilon A \cdot (t - ChangePoint)}\right]^{1/2}}{\left[1 - \dfrac{1}{\varepsilon A \cdot (t - ChangePoint)}\right]^{1/2}} - 1$$

These two principle causes of redshifts are depicted independently in Figure 14-7, below. Of course, the actual observed redshift is the sum of the two.

From the figure it is apparent that the Doppler-caused redshifts are quite minor until one is addressing light emitted only at times too early to be observable.

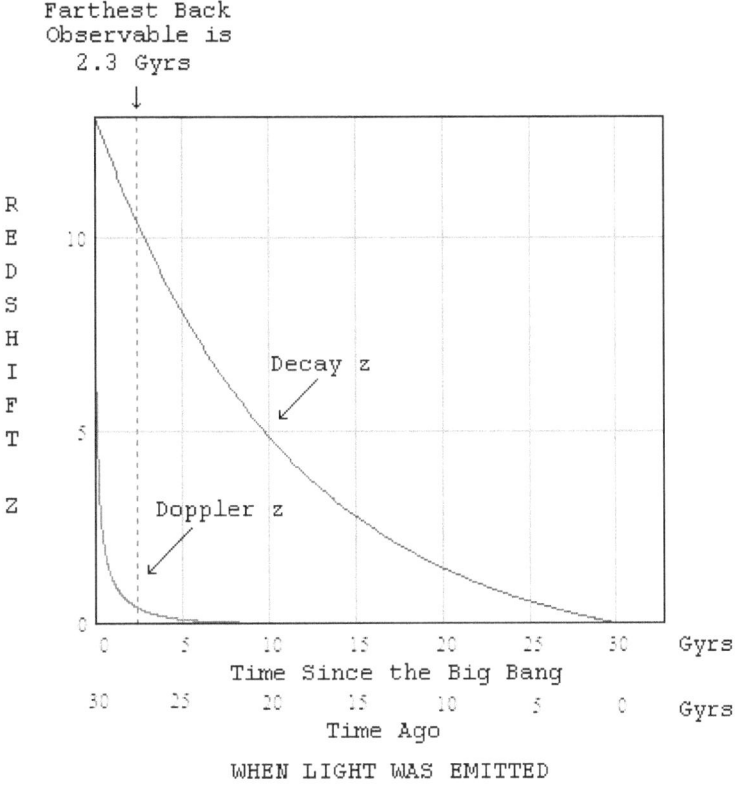

Figure 14-7
Redshifts: Caused by Universal Decay and by the Doppler Effect

The figure also makes clear why the age of the universe must be on the order of *30 Gyrs*. That amount of time is needed to include enough time constant periods, τ, that is *11.3373 Gyrs*, each to yield a maximum observable redshift [at *then = 2.3 Gyrs*] of $z > 10$ as in the figure. A small number of redshifts at *z = 10* have been reported with some indications of redshifts as high as *z = 12*. Improved instrumentation and techniques may well result shortly in confirmed detection of redshifts at $z > 10$. The value *age = 30 Gyrs* is an, at present, conservative best estimate taking into account currently known observational data.

3 - THE FATE OF THE UNIVERSE

Although the initial velocities of the "Big Bang" product particles were greater than the there / then escape velocity, as shown earlier above, their velocities now are all well under their escape velocity. We are used to escape velocity being escape velocity -- a simple yes or no proposition. The reason that that is not the case for the overall universe is the effect of the speed of light limitation.

The escape process is the conversion of kinetic energy into gravitational potential energy. If the initial kinetic energy is greater than the maximum possible gravitational potential energy then there will be escape. In the case of a rocket leaving Earth that process is accompanied by the rocket's velocity taking it farther enough away from the Earth that the gravitational effect is reduced in "proper" relation to the process. But the "Big Bang" product particles were not permitted to so travel, their actual velocity being limited to just under light speed as compared to the much larger theoretical non-relativistic velocity at which they would have had to have traveled outward for the accrued distance to correspondingly reduce the gravitational effect in "proper" relation to the process.

The actual velocities and the related escape velocities now at time $t = Age$ for the same distribution of initial "Big Bang" product energies analyzed earlier above is given in Table 14-9, below. As in the earlier analysis of the initial escape velocity, the present analysis is non-relativistic, using velocities greater than the speed of light rather than letting mass relativistically increase.

```
For:  Universe Age = 30 Gyrs, which means that:
      Initial Light Speed = 4.226,895,62 ·10^9 m/s
      Initial Gravitation Constant, G = 1.870,24 ·10^-7 m^3/kg-s^2
At Age = 30 Gyrs:
```

F-Factor	Outward from the Origin of the "Big Bang"*		
	Velocity[m/s]	Distance[G-Lt-Yrs]	Escape Velocity[m/s]
1	$0.00003814 \cdot 10^8$	2.157	$7.418 \cdot 10^8$
3	$0.0001505 \cdot 10^8$	2.400	$7.032 \cdot 10^8$
10	$0.0006230 \cdot 10^8$	2.727	$6.597 \cdot 10^8$
32	$0.002731 \cdot 10^8$	3.205	$6.085 \cdot 10^8$
55 Earth	$0.003700 \cdot 10^8$	3.374	$5.931 \cdot 10^8$
100	$0.01287 \cdot 10^8$	3.980	$5.461 \cdot 10^8$
316	$0.06673 \cdot 10^8$	5.388	$4.693 \cdot 10^8$
1000	$0.3974 \cdot 10^8$	8.034	$3.844 \cdot 10^8$

* = Decayed to Age

Table 14-9
Actual Velocities vs. Escape Velocities Now, at $t = Age$

One must immediately conclude that the entire material universe is ultimately destined to collapse back toward the location of its origin, just as a ball tossed straight up from the Earth's surface ultimately returns to its starting point. However, the case of the universe is more complicated than that of the simple ball and there are also two different considerations for the case of the universe: its matter and its radiation.

SECTION 14 - THE COSMOS' EXPANSION FROM THE ORIGIN TO THE PRESENT

a. *The Fate of the Universe's Radiation*

The fate of the radiation emitted from sources [primarily astral sources] throughout the universe is very different from the fate of the universe's matter. Most of the universe's radiation continues propagating outward forever, reduced in concentration inversely as the square of the distance from its source, and carrying outward in itself a significant amount of the universe's energy, which energy becomes essentially lost to the remainder of the universe, the universe's matter. That comes about as follows.

(1) *Gravitational Redshift and Light Escape*

When a particle of mass m climbs in a gravitational field its speed is reduced by the gravitation, which speed reduction reduces its kinetic energy, $\frac{1}{2} \cdot m \cdot v^2$. Conservation is maintained by the kinetic energy loss being replaced by gravitational potential energy increase.

A photon of frequency f has kinetic mass, m_{ph}, [even though it has no rest mass].

$$(14\text{-}41) \quad m_{ph} = \text{energy}/c^2$$
$$= h \cdot f / c^2$$

As light, with its kinetic mass, m_{ph}, climbs in a gravitational field, instead of its speed being reduced its frequency is shifted lower [toward the red]. The photon cannot slow down [to correspondingly reduce its kinetic energy as a particle of matter would] because it is constrained by its nature to only travel at light speed, c. Instead the photon frequency, f, decreases, which reduces its energy, $h \cdot f$, its energy of motion that corresponds to kinetic energy.

Then, for a photon to be able to escape from a gravitational field in a manner analogous to escape for a particle of mass, the photon energy, $h \cdot f$, must at least just exceed the depth of the gravitational potential energy pit, $G \cdot M \cdot m_{ph}/R$, that it experiences at the location where the photon is emitted. On that basis the calculation for photon escape would be that the photon frequency must be at least such that

$$(14\text{-}42) \quad h \cdot f_{minimum} > G \cdot M \cdot m_{ph}/R$$

however, the photon mass, m_{ph}, depends on f per equation 14-41 so that a directly solvable relationship cannot be obtained on that basis; photon escape is independent of photon frequency.

(2) *The "Schwarzschild Radius" and Escape*

Astrophysicists treat a quantity called the "Schwarzschild Radius". The line of thought is that the depth of a gravitational potential energy pit from which a particle must climb in order to escape is $G \cdot M \cdot m/R$ where G is the gravitation constant, M is the gravitating mass, m is the mass of the particle attempting escape, and R is the distance from the center of the gravitating mass at which the particle must begin its attempt. To escape, the particle's kinetic energy, $\frac{1}{2} \cdot m \cdot v^2$, must just exceed that potential energy so that, as presented earlier, the escape velocity is $v_{esc} = [2 \cdot G \cdot M / R]^{\frac{1}{2}}$.

From that formulation, as R decreases the required velocity, v increases. Therefore one can calculate a radius, R_S, the "Schwarzschild Radius", for any particular gravitating body mass, M, such that the required escape velocity, v_{esc}, is the speed of light, c, as follows.

(14-43) <u>For Light</u>, a Photon of Mass m_{ph}:

Photon Energy = Gravitational Potential Energy

$m_{ph} \cdot c^2 = G \cdot M \cdot m_{ph} / R = G \cdot M \cdot m_{ph} / R_S$

$R_S = G \cdot M / c^2$

(14-44) <u>For a Particle</u> of Mass "m":

Kinetic Energy = Gravitational Potential Energy

$\frac{1}{2} \cdot m \cdot v^2 = G \cdot M \cdot m / R$

$m \cdot c^2 = G \cdot M \cdot m / R_S$ [Because KE = TotalE - RestE, then as $v \to c$ TotalE >> RestE and KE $\to$ TotalE not ½·TotalE all because of the mass increase due to relativity.]

$R_S = G \cdot M / c^2$ [Solve for R_S]

[The usual presentation, that ignores the effect as in the above note, is $R_S = \underline{2} \cdot G \cdot M / c^2$]

No matter can travel at light speed, therefore matter located at or nearer to the center of the gravitating mass than R_S cannot ever escape. For radiation escape is independent of the frequency and depends only on the distance, R_S.

For the value of R_S for the universe at the instant of the "Big Bang":

· from equation 14-16 G was $G(0) = 1.870 \cdot 10^{-7}$ m³/kg-s²,

· from equation 14-13 M was $M_{Universe} = 3 \cdot 10^{49}$ kg, and

· from equation 14-3 c was $c(0) = 4.226 \cdot 10^9$ meters/sec.

Then, the value of R_S for the universe at the instant of the "Big Bang" was $3.14 \cdot 10^{23}$ meters which is [0.033 G-Lt-Yrs] and at that instant the actual distance from the center of the "Big Bang" was much less, $d_0 = 4.0 \cdot 10^7$ meters. Therefore, at that time, $t = 0$, no matter nor light could escape from the "Big Bang" as already demonstrated and summarized for matter in Table 14-9, above. The inability of matter to escape did not change thereafter.

SECTION 14 - THE COSMOS' EXPANSION FROM THE ORIGIN TO THE PRESENT

However, in its rapid initial expansion at a speed of very nearly $c(0)$, after time approximately $[R_S \div c(0)]$, that is the first two to three million years, the light-source matter of the universe had moved out from the origin to beyond the "Schwarzschild Radius" and whatever radiation was emitted thereafter was free to travel outward forever.

b. The Fate of the Universe's Matter

The universe's matter, however, was already embedded in the impossibility of escape and it only remains to investigate its fate.

To this point the material presented has consisted of analytical deductions and reasonable estimates based on fundamentals of physics, the available data, and the tenets of the theories involved. Now, with regard to the fate of the universe's matter, some of what is presented must be limited to "educated" speculation as to the implied future while some still remains reasonable analytical deductions.

Clearly the large range of the present velocities of the universe's matter and of its varied present distances outward from the origin per Table 14-9 means that the universe's matter's gradual slowing - direction reversal - inward collapse will result in a wide range of arrival times at the origin of the original expansion of the universe's various portions. [That as juxtaposed to the concept of a universe all together collapsing and then re-exploding outward in a succession of "big bangs" as has been hypothesized in the not too distant past.] There are, then, several possibilities to be considered.

- Matter so arriving at the initial origin crashing into like kind so arriving matter.

    Recently there have been analyses of what happens when a large asteroid crashes into the Earth, the energies involved and the resulting destruction being immense. One can only [speculatively] increase those energies and their results many orders of magnitude to conceive of what would happen at the collision of two planetary bodies, two suns, or two galaxies.

    However, the collision would be kinetic and produce great heat, breakdown into particles, and great kinetic energy of those product particles. It would not be as a nuclear fission nor fusion explosion, that is it most likely would not involve a major conversion of matter to energy.

- Matter so arriving at the initial origin encountering there nothing but empty space.

    Unlike the case of the ball tossed upward from the Earth's surface in which case the Earth is still there when the ball falls back down, it would seem that there is now likely nothing but empty space at the location of the initial origin. A portion of the universe's matter arriving there unopposed would be traveling at high speed [most likely the same [then outward but now inward] speed as was imparted to it in the original "Big Bang" [but as reduced by the Universal Decay of the speed of light]. That body of matter would pass on through and proceed outward again in its own "personal" replay of its earlier role.

Except, that is, that the first time the gravitating matter of the universe was initially all concentrated at the origin whereas the second time that matter is scattered over a large universe volume. The gravitational conditions would be different for the second pass and the escape velocity would also be different. One can only [speculatively] imagine various scenarios for the further travel of that portion of the universe's matter and its peers / partners.

- Matter so arriving at the initial origin and there encountering anti-matter.

There are two alternative hypotheses with regard to anti-matter creation in the "Big Bang":

· Anti-matter was created, but in a lesser amount than ordinary-matter, and quite shortly thereafter all of the anti-matter mutually annihilated with an equal amount of ordinary-matter leaving essentially no remaining anti-matter and a small remaining amount of ordinary-matter, which is the matter of our universe. In this hypothesis there is no, or negligible anti-matter in today's universe.

This alternative voids the "matter arriving encountering anti-matter" possibility.

· Matter and anti-matter were created in equal, "mirror" amounts and, while some of it promptly mutually annihilated, equal amounts of each participated in the outward expansion quickly enough to survive. Thus our universe has matter portions and similar anti-matter portions and their continued separation in space largely preserves their continued independent existence.

In this hypothesis matter so arriving at the initial origin could encounter anti-matter, which would result in a mutual annihilation. Unlike the kinetic collision case, the result would be an immense amount of energy radiated mostly as gamma rays.

With regard to those two alternative hypotheses the last Reference, *The Problem of Big Bang Matter vs. Antimatter Symmetry* favors the second of the two.

SECTION 14 - THE COSMOS' EXPANSION FROM THE ORIGIN TO THE PRESENT

The behavior of anti-matter is such that there is no way to discriminate whether a distant astral source is matter or anti-matter: the gravitation is the same; the light emitted is the same.

c. *The Ultimate End of the Universe.*

(1) The Universal Decay Will Continue

The universe will continue shrinking to beyond the point of extremely minute, all to no noticeable effect on its internal functioning no matter how small it becomes relative to the size that it is now or originally was.

If one looks back one million years ago, lengths then were greater than the corresponding lengths today by a factor of 1.0009. Clearly the universal decay has little significance in day to day life. In fact, its only significance is for astronomers, because only they can look back into the past far enough to see the effects of the extremely slow decay.

Everything decays proportionately. The ratio at any time, now or in the past or in the future, of the size of things relative to things does not change at all. There is no fixed objective reference by which one could appreciate or notice the decay other than those accessible only to astronomy. Everything is shrinking, but to no noticeable effect. Whatever happens to be left of the universe some inconceivable number of aeons from now will be so extremely minute compared to the size of things in today's universe as to seem to us as nothing. Yet it will operate, function, behave according to the same rules as our universe now, as if it had not decayed at all [again except astronomically], but subject to the events below.

(2) The Universe's Matter Will Gradually Completely Obliterate

Whatever time it takes, eventually all of the universe's matter will be obliterated in mutual annihilations. The process will be a kind of universe "Russian Roulette", annihilations depending randomly on the simultaneous arrival of matter and anti-matter portions of the universe at the location of the initial origin. Such annihilations will extend only to the extent of arriving masses being equal; the un-annihilated surplus of the greater being hurled outward again for another excursion and later chance of annihilation upon its return.

(3) The Universe's Radiation and Energy Will Be Dispersed in Endless Space

All of the radiation and energy of the matter annihilations along with all of the astral and other radiation and energy from the beginning on [including radiation absorbed and subsequently re-radiated] will disperse outward in space, gradually reddening and so reduced by inverse square dispersion as to eventually amount to essentially nothing.

(4) Nothing to Nothing ...

In the same way as for we humans when our span of life ends it is said, "Ashes to ashes and dust to dust ....", so for the universe it can be said, "It came from nothing and eventually passes on to nothing, to that from which it came".

– End –

References

[1] Zwicky, F. (1933), *Die Rotverschiebung von extragalaktischen Nebeln*, Helvetica Physica Acta **6**: 110–127, this article may be found at http://adsabs.harvard.edu/cgi-bin/nph-bib_query?bibcode=1933AcHPh...6..110Z.

[2] J. D. Anderson, P. A. Laing, E. L. Lau, A. S. Liu, M. M. Nieto, and S. G. Turyshev, *Indication, from Pioneer 10/11, Galileo, and Ulysses Data, of an Apparent Anomalous, Weak, Long-Range Acceleration*, Phys. Rev. Lett. **81**, 2858 (1998).

[3] John D. Anderson, James K. Campbell, John E. Ekelund, Jordan Ellis, and James F. Jordan, *Anomalous Orbital-Energy Changes Observed during Spacecraft Flybys of Earth*, Physical Review Letters, PRL **100**, 091102 (2008).

[4] A. Kashlinsky, F. Atrio-Barandela, D. Kocevski, H. Ebeling, *A Measurement of Large-Scale Peculiar Velocities of Clusters of Galaxies: Results and Cosmological Implications*, Astrophysical Journal Letters, Print edition October 20, 2008; online week of September 22, 2008.

[5] R. Ellman, *The Origin and Its Meaning*, The-Origin Foundation, Inc., http://www.The-Origin.org, 1996. [The book may be downloaded in .pdf files from http://www.The-Origin.org/download.htm].

[6] R. Ellman, *Analysis of the "Big Bang" and the Resulting Outward Cosmic Expansion: Hubble - Einstein Cosmology vs. The Universal Exponential Decay*, available at http://www.arXiv.org, arXiv:physics/0004053 [pdf].

[7] A. Kashlinsky, F. Atrio-Barandela, H. Ebeling, A. Edge, and D. Kocevski. *A New Measurement of the Bulk Flow of X-Ray Luminous Clusters of Galaxies*. The Astrophysical Journal, 2010; 712 (1): L81 DOI: 10.1088/2041-8205/712/1/L81

[8] R. Ellman, *The Problem of Big Bang Matter vs. AntiMatter Symmetry*, available at http://www.arXiv.org, arXiv:physics/0007058 [pdf].

www.ingramcontent.com/pod-product-compliance
Lightning Source LLC
Chambersburg PA
CBHW062217220526
45471CB00009B/3245